What figures do the constellations of the night sky represent? What are the properties of the stars that they comprise? And where in the sky can they be found? This unique reference gathers together more information on the constellations than any other single work to date.

The constellations can be readily compared, and a general view of them developed, using the tables that make up the first part of the book. These provide a wealth of information, covering all the essential properties of the constellations. In the second part of the book, each constellation is taken in turn, with a star chart and map illustrating the associated celestial figure, supported by a comprehensive list of essential properties.

This highly illustrated volume provides the most complete reference to date covering all factual aspects of the constellations – for astronomers, both amateur and professional, educators and science writers.

DATE DUE

			PRINTED IN U.S.A.

THE CAMBRIDGE GUIDE TO THE CONSTELLATIONS

The Cambridge Guide to the Constellations

MICHAEL E. BAKICH

Planetarium Director, Kansas City Museum

CAMBRIDGE
UNIVERSITY PRESS

ess Syndicate of the University of Cambridge
rumpington Street, Cambridge CB2 1RP
New York, NY 10011-4211, USA
Oakleigh, Melbourne 3166, Australia

© Cambridge University Press 1995

First published 1995

Printed in Great Britain at the University Press, Cambridge

A catalogue record for this book is available from the British Library

Library of Congress cataloguing in publication data

Bakich, Michael Eli.
The Cambridge Guide to the Constellations / Michael Eli Bakich.
   p.   cm.
ISBN 0 521 46520 6 – ISBN 0 521 44921 9 (pbk.)
1. Constellations–Observers' manuals.   2. Astronomy–Observers' manuals.   I. Title.
QB802.B35   1995
523.8'022'3–dc20   94-4678 CIP

ISBN 0 521 46520 6 hardback
ISBN 0 521 44921 9 paperback

TAG

Contents

Preface

My love for the constellations that populate the nighttime sky began when I was seven years old, during the summer before my third grade in elementary school. One day during that formative period of my life, my parents presented me with a small, boxed set of flash cards. On the front side of each card was a pattern of stars – a constellation. Connecting the brighter members of the group was a set of straight lines, giving the whole the appearance of a figure of some sort.

The reverse of the card held the name of the constellation, the meaning, and a list of the bright stars contained therein. The intended purpose was for one person (my mother, as it soon turned out) to hold up a card, and for a second person (me) to guess the constellation.

I simply devoured the information. I became a walking "expert" on constellations, at least the ones pictured on the cards. Of course, this information was limited to what the cards contained. No matter. At the tender age of (almost) eight, I began proselytizing anyone who would listen. My life's vocation was established, even before I experienced the real stars.

At one point, my mother, who had patiently endured the exhuberant onslaught far longer than I, looking back, could have expected, gently suggested that I apply my newly-found awareness to the real sky. That night, under cover of darkness, a breakthrough was achieved. Memory fails as to which constellation was the first to be identified. Possibly Boötes led the way, or maybe Leo, or perhaps Ursa Major holds the distinction. At any rate, the transference to 'real life' had been made.

The next discovery – that my humble set of flash cards could not account for all the stars I observed in the night sky – was to set a pattern of questioning that has continued unabated to this day. Were all the constellations represented? Did some stars in the sky not belong to any constellation? Was I seeing the entire sky? If Sirius was the brightest nighttime star, which star was second? All these questions, and many more, could be summed up in one: where do I look for more information? Hopefully the writing of this book will break this pattern.

It is my intention to include within this book *all* pertinent (and much, how shall we say, 'frivolous') information about every constellation in the sky. I fully realize that any undertaking of this magnitude is doomed from the start. The very fact that this book is not three times its present size is proof. Yet, I felt compelled to try.

I have often been called upon to write articles and descriptions about certain constellations, I have presented thousands of 'sky lectures' under the domes of planetaria of all sizes, and I have written full-length planetarium programs describing the sky in a monthly or seasonal time frame. These tasks have necessitated an almost continuous search for facts related to the constellations. How does Scorpius rank in size? (One reference). Which constellations border Scorpius? (A second reference.) On what date is Scorpius in conjunction with the sun? (A third reference.) And on and on it went. I have amassed quite a library trying to answer such questions.

'Why,' I have often wondered, 'has no one compiled *one* reference work with all, or nearly all, the information I need?' The present work was undertaken to fill this gap. It is a gap that, in my estimation, has existed for centuries.

Acknowledgements

First and foremost, I would like to thank my very dear friend Ray Shubinski. Lacking Ray's gentle encouragement, this project never would have gotten off the ground. Without Edmund Halley, the law of gravitation would have lain unannounced in Newton's study, and Newton himself might only have been a historical footnote. Thanks, Ray, for playing the part of Halley and for being the spark that got this fire going.

Another motivational award goes to Rita Mortenson. Rita was a source of endless optimism and an example to which I looked during the difficult middle stages of writing.

Professionally, I am very grateful to Bruce Bradley of the Linda Hall Library in Kansas City, Missouri. Bruce's tireless efforts and vast store of knowledge made this endeavor much easier than it would have otherwise been. The fabulous collection of the Linda Hall Library was the source of eighty-five of the eighty-eight ancient star maps used in this work.

The other three maps (of Carina, Puppis, and Vela) came from the British Library in London, UK. I would like to thank the Map Librarian and the staff of the British Library for providing these photographs from a long way off.

All of the modern constellation charts were computer-generated using The Sky astronomy software for Windows from Software Bisque, Golden, Colorado. Thanks to Matthew L. Bisque for providing this software. For their computer expertize I would like to thank Kevin Wehner and John Casey, with extra thanks to my friend John for the use of his incredibly fast computer with which I generated all star charts in this book.

I would also like to take this opportunity to recognize the team at Cambridge University Press. First, thanks to Simon Mitton for believing, as I did, that this work should be shared, for counseling patience, and for an early morning call that really made my day. Second, thanks to Adam Black for his supervision and committment to communication during the project. And a very large thanks to Jo Clegg whose meticulous editing of the copy has made this a much better book than when it was submitted.

Finally, I would like to thank all the family and friends whose high hopes and countless prayers made this effort both possible and worthwhile.

Figure acknowledgements

All but three of the antique star maps are reproduced courtesy of the Linda Hall Library, Kansas City, MO. Photographs by the author.

The exceptions are maps of Carina, Puppis, and Vela, which are reproduced by permission of the British Library, London. Photographs by the British Library.

1 Lists

Explanations for lists

(The lists are presented in alphabetical order.)

Alphabetical list of the constellations (Pages 16–17)

This list provides an alphabetized inventory of the names of the eighty-eight modern constellations. Next to each name is the figure represented by the constellation.

Abbreviations of constellation names (Pages 18–19)

The three-letter abbreviations listed were first proposed by Ejnar Hertzsprung and Henry Norris Russell at the First General Assembly of the International Astronomical Union held in Rome in 1922. (*Transactions of the IAU*, vol. 1, edited by A. Fowler.) Note that in two cases (those of Hydrus and Sagitta) a letter appears in the three-letter abbreviation that is not found in the name of the constellation! This apparent mistake is explained by the fact that the possessive (or, genitive) form of each constellation name was used as the basis for the abbreviations. See the explanation under 'Possessive forms of the constellation names' for more information.

Asterisms within the constellations (Pages 20–23)

In the sky lore of some ancient cultures, asterisms were looked upon as co-equal with constellations. In fact, *Webster's New World Dictionary*, 1965 edition, defines the term 'asterism' as (1) a constellation; and (2) a star cluster. It may be well to note that, in many cases throughout history, and even now, certain asterisms have been much easier to identify than most of the constellations.

In the same vein, Minsheu, in his *Dictionary* of 1625, defines an asterism as a 'configuration of fixed stars, an imaginary form devised by the astrologers, the better to conceive and distinguish asunder the fixed stars...' Obviously, this definition could apply equally well to constellations.

Asterisms are now defined as 'unofficially' recognized groupings of stars. Sometimes an asterism may be formed from the stars of a single constellation, as in the case of the Big Dipper in Ursa Major. Other times, asterisms are formed using stars from several constellations, as in the case of the Winter Triangle, which is formed using the α stars from the constellations Orion, Canis Major, and Canis Minor.

Biblical references to the constellations and stars (Pages 24–25)

This list is included because of the tremendous significance of the Bible to western civilization. It is interesting that, in the entire Bible, there are only three specific references to constellations. In addition to these, all other astronomical references are noted. Most of these, it must be stated, refer to either 'Heaven and Earth' or to the Sun or Moon.

Bordering constellations (Pages 26–30)

When describing a particular constellation, it is often useful to know which constellations lie on its borders. For those unfamiliar with certain constellations, this can be helpful in directing them to the proper area of the sky.

In this list the letters in parentheses give the direction *from* the main constellation. For example, one would say that Andromeda is bordered on the north by Cassiopeia, on the west by Lacerta, etc. For large constellations which border smaller ones, or for the many constellations which wind about the sky, more than one bordering direction is possible (see, for example, Crux or Serpens).

The 200 brightest stars (Pages 31–35)

This list provides data for the 200 brightest stars in our sky, measured by their apparent visual magnitude. Magnitudes were obtained from the fourth edition of the *Yale University Observatory Bright Star Catalogue* by Dorritt Hoffleit. In column 1 we find the Bayer designation of the star; column 2 gives the most common form of the name of the star (if any); column 3 is the Yale Bright Star number. Where two numbers appear (as for 3 α Cen, 5459/60) they represent a double star where the listed magnitude is the combined magnitude of both components; column 4 lists the apparent magnitude to two decimal places as given in the *Bright Star Catalogue*; column 5 gives the approximate absolute magnitude of the star; and column 6 gives the approximate distance to the star in light years. For nearby stars, measures of parallaxes provide a reasonably accurate way to determine distances.

$$\text{Distance (in parsecs)} = 1 / \text{Parallax (in arcseconds)}$$

Many other sources were consulted to determine approximate distances to remote stars.

Central points of the constellations (Pages 36–38)

For those who need a quick reference giving the location of the constellations, this list defines the mid-point of each. Two sets of numbers are given. The 'average' central point, for those occasions which do not require precise coordinates, is accurate only to the nearest hour of right ascension and the nearest five degrees of declination. The exact listing is accurate to one minute of right ascension and one half degree of declination.

The numbers given are the mid-points in right ascension and declination between the directional extremes. These numbers are not the 'geo-celestial-graphical' center of the figures. That is, the coordinates given by the two numbers would not be the so-called 'balance point' of the constellation.

Directional extremes (Pages 39–41)

This list provides the boundary limits of the constellations in each of the four principal directions. Such a list is useful for a number of reasons. To those observers familiar with the sky, it gives an idea as to when a constellation will begin to rise (or set) on a particular date of the year.

For example, on or around 20 April, when will the constellation of Cygnus begin to rise, approximately? First, calculate the approximate right ascension of the Sun. On 21 March, which is the vernal equinox, the Sun's right ascension is 0h. In a month, that number will increase by 2h, so the sun's right ascension on April 20 is 2h. At sunset, then, a point opposite the Sun (the 'anti-solar' point, 180° across the sky in the east) will have a right

4

ascension of 2h + 12h, or 14h, approximately. Checking the list, we see that the western directional extreme of Cygnus is 19h07m. We use the value for the *western* extreme of the constellation because that is the first part of Cygnus which will become visible in the east. Simple subtraction shows that there is a difference of 5h07m between the anti-solar point at sunset and the first appearance of Cygnus. Therefore, the western edge of Cygnus will rise approximately 5 hours after sunset on the given date.

Also, we note that the eastern directional extreme of Cygnus has a right ascension of 22h01m. This means that the entire constellation of Cygnus will have fully risen 2 hours 54 minutes (22h01m – 19h07m) later, or approximately 8 hours after sunset on the given date.

Extinct constellations (Pages 42–48)   Astronomers now recognize 88 constellations. These constellations cover the entire sky with no overlaps and with no gaps between them. This, however, is a fairly recent development. Prior to 1928, celestial mapmakers were free to populate the skies as they pleased, with only the mildest of restraint placed upon them by other astronomers and cartographers. The very fact that many of these star groupings have become extinct compels one to quote Darwin's axiom of the 'survival of the fittest.'

The first star maps and globes had less than 50 constellations, yet despite that relatively small number some of those have not survived to the present day. The most notable example is the truly massive constellation of Argo Navis, the legendary ship which carried Jason and the Argonauts in search of the golden fleece. Argo Navis is also probably the only example of a constellation that fell into disuse for purely practical reasons: it was simply too big. So, in the middle of the eighteenth century, it was decided that it would be subdivided into the constellations of Carina, Puppis, and Vela. This has made the cataloging of objects within the boundaries of these star groups much easier for modern-day astronomers. So ingrained was the use of the designations for the stars of Argo that, when this constellation was rent asunder, the original set of Bayer letters was retained, with no additional ones being added. Thus, Puppis and Vela have no alpha star; the single alpha star of Argo Navis (Canopus) was given to the constellation of Carina.

Almost every other constellation which has gone the way of Argo Navis has vanished because of non-acceptance by the community of astronomers at large. Many of these star groups were named in honor of the patron or monarch of the astronomer. Needless to say, outside of the boundaries of the relevant country these tributes were not well received. A similar situation occurred in 1781, when a row developed after Sir William Herschel attempted to name his newly-discovered planet 'Georgium Sidus' or 'George's Star.'

This is a list of 'extinct' constellations, the (current) designations of the stars which were used to form them, the figure represented by the constellation, and the astronomer/cartographer whose map or globe first mentioned the figure, and the year. Those constellations which were mentioned on only one map or globe, thus achieving no degree of notoriety whatsoever, are either ignored or only briefly noted.

The magnitude system **(Pages 49–50)**

The first observer to describe and catalog differences in brightnesses of stars was the Greek astronomer Hipparchus, who lived in the second century BC. He divided his listing of approximately 850 visible stars into six brightness ranges, or magnitudes. The brightest he classed as stars of the 1st magnitude, and the faintest as stars of the sixth magnitude. His system was used, almost unchanged, for more than 1800 years.

After the invention (and subsequent upgrading) of the telescope, it was deemed necessary to expand this system. Many stars fainter than those listed as 6th magnitude by Hipparchus were now visible. In addition, it was noted that stars of 1st magnitude varied greatly in brightness. Around the time of the great observational astronomer Sir William Herschel, a loose system was adopted that defined two stars differing by one magnitude as having a brightness difference of approximately two and a half.

William R. Dawes (1799–1868) proposed, in 1851, a simple and effective method of photometric comparison, depending upon the principle of equalization by limiting apertures, to the problem of a fixed standard of stellar magnitude. (*Monthly Notices*, Royal Astron. Soc., vol. xi. 187.)

In 1856, Norman R. Pogson suggested that all observations be calibrated by using the constant $10^{2/5}$. The ratio between magnitudes thus becomes approximately 2.5118865. That is, a star of a given magnitude is 2.5118865 times brighter than a star one magnitude fainter. At this time, the concept of using magnitudes equal to and less than zero also came into being. The rationale involved keeping some semblance of the original system, where the general limiting magnitude of the human eye was approximately 6th magnitude. With this limitation and Pogson's mathematical formula, it became evident that the brightest stars were much brigher than 1st magnitude, to say nothing of the bright planets, the Moon, and, of course, the Sun.

The number (2.5118865) is the fifth root of 100. Therefore, a difference of five magnitudes is equal to a 100-fold difference in actual brightness. Thus Sirius (α CMa) at magnitude –1.46 is 100 times as bright as Wasat (δ Gem) at magnitude +3.53.

In the given table, if values are sought which are not listed, simply multiply the ratios of the magnitude differences which, when added, give the desired difference. For example, to find the brightness ratio between Antares (α Sco) at magnitude 1.2 and Ras Algethi (α Her) at magnitude 3.5 (difference = 2.3), simply multiply the ratios of 6.3095738 (for a magnitude difference of 2.0) and 1.3182567 (for a magnitude difference of 0.3). Thus, Antares is $6.3095738 \times 1.3182567 = 8.3176379$ times as bright as Ras Algethi, approximately.

Adding magnitudes involves the use of a simple formula (from *Astronomical Formulae for Calculators*, Jean Meeus, Willmann–Bell, Inc., Richmond, VA., 1979):

$$m_c = m_2 - 2.5\log(10^x + 1)$$

Where: m_c is the combined magnitude of the system
$x = 0.4\,(m_2 - m_1)$
m_1 and m_2 are the magnitudes of the stars

Exact magnitudes for the 200 brightest stars are given within this list. Exact magnitudes are also provided in the list of navigational stars. Magnitudes are also given in the lists of the 200 nearest stars, stars with the largest proper motion, and star names. The magnitudes of the stars listed in these three lists have been rounded to the nearest 0.1 magnitude.

Messier objects (Pages 51–57)

Charles Messier (1730–1817) was a famous comet hunter who compiled a list of non-stellar objects that, in the small telescopes of his day, could have easily been mistaken for comets. Messier's original list consisted of 103 objects and was published in 1784, only three years after the discovery of the planet Uranus, by Sir William Herschel. Several additional objects were added in later years, and present-day amateur astronomers usually recognize 110 objects, although a few are questionable.

Three lists under this heading are provided. The first gives the Messier objects in numerical order; the second arranges them according to constellation; and the third catalogs them by right ascension.

Meteor showers (Pages 58–61)

Apart from individual stars, nothing in the heavens is more characteristically associated with constellations than meteor showers. Most showers are named after the constellation in which their radiants are located. When more than one shower occurs within the boundaries of a particular constellation, the location of the individual radiants may be differentiated by associating them with certain stars. Even in such cases, however, the proper name of the star is never used, only the Bayer designation with the genitive form of the constellation's name. Thus, for example, there is no meteor shower called the 'Antarids.' Rather, it is referred to as the 'α Scorpiids'.

For meteor showers which have been discovered more recently, association with individual stars has become common even if there is no shower named for the constellation as a whole, as with the δ Cancrids. Meteor showers having two radiants are due to parallel streams of particles. These streams are caused by the breakup of the comet from which the shower originated.

Careful inspection may reveal that some of the shower radiants are outside the boundaries of the constellations for which they are named. The reason for this is that meteor showers are not instantaneous events. It may take several weeks for the Earth to pass through the stream of particles which contribute to the shower. As the Earth (in its orbit around the Sun) moves through the stream, the position of the radiant changes. In the list, the coordinates of the radiant are given only for the date of maximum. It is entirely possible, then, that the coordinates may lie slightly outside the official constellation boundary.

The first of the lists under this heading is arranged chronologically. The second is by constellation. Along with the approximate date of the maximum strength of the shower and the shower name, the coordinates of the radiant and the zenithal hourly rate (ZHR) of the shower have been provided. The ZHR is used to express the relative strength of the shower. It is a

7

ficticious number an expert observer would see under a dark sky (having limiting magnitude +6.5) with the radiant directly overhead. It is uncommon, therefore, for an observer to match or exceed the ZHR of any particular shower.

Midnight culmination dates of the constellations (Pages 62–64)

By definition, culmination is the passage of a celestial body across an observer's meridian. Usually, and especially in most visual applications, this refers to 'upper' culmination, or the crossing of the object closer to the observer's zenith. A 'lower' culmination of an object is also possible, 180° from 'upper' culmination. For the majority of objects, this is invisible. The only exception occurs when the object is circumpolar for a particular location. Upper culmination is sometimes referred to as the 'transit' of the object across the meridian.

'Midnight' culmination refers here to the meridian passage of a constellation (more specifically, the *center* of the constellation) when the constellation lies 180° from the Sun. Thus, the constellation culminates at midnight, which, incidentally, refers to exact *mid*-night, and may not be the same as 24:00, especially if daylight saving time is in effect. It may be instructive to note that when a constellation culminates at midnight, at the same moment the Sun is at lower culmination.

Another way to think about the midnight culmination date of a constellation is as the date when that constellation rises at sunset, sets at sunrise, and is visible throughout the entire night. It is, therefore, the best night of the year to view a particular constellation, depending, of course, on the phase of the Moon.

The Moon and the planets in the constellations (Page 65)

This list is provided to demonstrate that the Moon and planets appear within the boundaries of many constellations, not just the twelve traditional constellations of the zodiac. For example, on 30 Jan 1988, at 20h UT, the Moon's position lay within the constellation of Auriga. On 20 Jun 1981, Mercury was 'in' Orion, and in April, 1980 Pluto was 'in' Boötes. Many other cases could be cited. The explanation is that the orbits of the Moon and planets (especially Pluto) are inclined to the plane of the ecliptic, that is, to the Earth's orbital plane. The Moon can vary as much as 5° above or below the ecliptic, Mercury about $4\frac{1}{2}°$, and Pluto more than 17°.

Names of the constellations around the world (Pages 66–71)

Today, all countries recognize the 'Latin' names of the constellations, as given in the initial list of this book. Almost all published research uses those names and their Latin genitive forms. However, it is interesting to see the names of the constellations in several foreign languages. As can be seen, in many cases the names are often not proper names (as in English), but rather they refer to the figure represented by the constellation.

The navigational stars (Pages 72–74)

The first official nautical ephemeris or almanac was published in France in 1687. In 1852, the United States Navy's Depot of Charts and Instruments published the first *American Ephemeris and Nautical Almanac* for the year 1855. This volume has been published every year since then.

The Nautical Almanac originally tabulated data on a total of 173 stars suitable for use in celestial navigation. Of these, only 58 were normally used; these are the 57 so-called 'selected stars,' or 'navigational stars,' plus Polaris. It is these stars that are given in this list. Currently, *The Astronomical Almanac* provides data on a total of nearly 1500 'Bright Stars.'

The 200 nearest stars (Pages 75–81)

This list provides data for the 200 nearest star systems in our sky, measured by their parallax. The Sun is included as a reference, but is not numbered. In column 1 we find the designation of the star; column 2 gives the constellation in which the star is located; column 3 is the apparent magnitude; column 4 lists the absolute magnitude; column 5 gives the parallax of the star in arc seconds; and column 6 gives the approximate distance to the star in light years. Distances to the stars were calculated directly from the parallaxes using the formula

$$\text{Distance (in parsecs)} = 1 / \text{Parallax (in arcseconds)}.$$

This result was then multiplied by 3.26 to give the distance in light years. The main source consulted was the Royal Observatory Annals, Number 5, *Catalogue of Stars within Twenty-Five Parsecs of the Sun*, by Sir Richard Woolley, Elizabeth A. Epps, Margaret J. Penston, and Susan B. Pocock, Herstmonceux: Royal Greenwich Observatory 1970.

The 'new' constellations (Page 82)

Slightly more than half of the constellations in use today originated in antiquity. This is a list of the 'recent' additions. These star groups have been adopted by astronomers to give us the official 88 constellations discussed in this work. The following is a brief sketch of the astronomers who introduced the 'new' constellations:

Johannes Hevelius (1611–1687)

A German astronomer best known for his studies of lunar features. In 1687, he completed a celestial atlas of 56 engraved maps which accompanied a star catalog. Within this atlas were presented seven new constellations. Hevelius also studied the Sun's rotation, named the bright areas of the solar photosphere 'faculae,' and was the first to observe the phases of Mercury.

Frederick de Houtman (1540–1627)

A Dutch navigator who sailed with Keyser (see below) to the East Indies in 1595–7. Upon his return, he forwarded all observations to Plancius, who distributed the information about the new constellations to astronomers and cartographers of the day.

Pieter Dirksz Keyser (d. 1596)

A Dutch navigator and one-time student of Plancius. From 1595 to 1596, Keyser was aboard one of four ships which sailed to the East Indies. Along with de Houtman (see above), and working with instructions from Plancius, Keyser mapped nearly 200 stars near the south celestial pole. The two navigators introduced 12 'new' constellations, however, one of these,

Triangulum Australe, had already been noted by Vespucci as early as 1503. Keyser died off Java before the voyage was completed.

Nicolas Louis de Lacaille (1713-1762)

A French astronomer who cataloged the positions of tens of thousands of stars. From 1750–3, Lacaille was in South Africa, studying the stars of the southern hemisphere. Upon his return, he published a star catalog and map. Lacaille invented 14 new constellations to fill the massive voids in the southern sky. Lacaille did not imitate ancient celestial cartographers, but chose constellations which represented contemporary scientific equipment. Lacaille's importance in the astronomical community of his day was demonstrated by the fact that although his star patterns were non-traditional, they were immediately accepted by astronomers and map makers.

Gerard Mercator (1512–1594)

A Dutch mapmaker, probably the most famous of all. Mercator's style was widely copied, although his lifetime output was small, as he drew and engraved all his maps. Mercator did not originate the style of mapmaking now known as 'Mercator's Projection,' but he was the first to apply it to navigational charts. He also made mathematical and astronomical instruments and globes.

Petrus Plancius (1552–1622)

A Flemish mapmaker and theologian who settled in Amsterdam in 1585, because of religious persecution. He was the first cartographer of the Dutch East India Company (1602–19), and energetically promoted plans for Dutch trade to the East. During his lifetime, Plancius produced over 100 individual maps and a few globes, but he never published an atlas. A mapmaker of the highest order, Plancius was considered second only to Mercator during his lifetime.

Amerigo Vespucci (1451–1512)

An Italian navigator after whom America is named. In 1503, Vespucci was the first to depict the constellations of Crux and Triangulum Australe in a letter to Pier Lorenzo de Medici.

The 'original' 48 constellations (Pages 83–84) When one mentions the original constellations of the Greeks, one generally refers to those which appeared on the uranographic globe of Eudoxus of Knidos (c. 403 BC to c. 350 BC). Unfortunately, no work of Eudoxus survives. Most of his star figures (44) were later included, c. 270 BC, in the *Phainomena*, by the poet Aratos, who lived in the court of the King of Macedonia, Antigonas Gonatas. It is these constellations which are listed.

Other ancient chroniclers of the sky were Eratosthenes (c. 276 BC to c. 195 BC), who mentioned 42 constellations, and Hipparchus (c. 146 BC to c. 127 BC), who listed 49 in his *Catalog* and 46 in his *Commentary on Eudoxus and*

Aratos. Homer, writing in the *Iliad* and the *Odyssey*, only mentions five individual stellar groups: the Clusterers (Pleiades), the Bear (Ursa Major), Orion, the Ploughman (Boötes), and the Rainy-ones (Hyades). Homer does, however, make references to almost every figure which formed the primitive constellations (the only exceptions being the Crab, the Crow, and the Scorpion), and was seemingly quite familiar with them when these two major works were written, certainly before 750 BC, and probably sometime before 1000 BC.

Presently, the constellations usually thought of as the 'original' constellations of the Greeks are the 48 mentioned by Ptolemy (AD 73 to AD 151) in the seventh and eighth books of his *Almagest*. Ptolemy's list is given after that of Eudoxus. It is almost certain that this 'star catalog' of Ptolemy was the same one given by Hipparchus, edited by Ptolemy to reflect the changes in precession between his time and that of Hipparchus.

The constellations listed by Ptolemy are chosen due to the tremendous influence the writings of Ptolemy had on how astronomy was understood, this influence lasting more than a dozen centuries after his death. His star catalog summed up the results of the star-gazers of early Greece and Western Asia. It also supplied the foundation for the efforts of those which followed him, including the great schools of Arabian and Medieval astronomers. The manner in which Ptolemy himself entitled his list implied that it was no novelty, but the 'Authorized Version' (canon) of the constellations and their stars.

It may be noted that of these 'original' constellations, only one, Argo Navis, is no longer recognized. That unwieldy constellation (although still often illustrated as a complete ship) was subdivided by a number of ancient celestial cartographers beginning in the eighteenth century. It was dealt its death blow by the International Astronomical Union in 1928, and now more manageably resides as the three constellations of Carina, Puppis, and Vela.

Overall brightness of the constellations (Pages 85–89)

For the purpose of this list, we define the 'overall brightness' of a constellation as the number of visible stars per unit area. To facilitate easy comparison, and to avoid using extremely small numbers, the unit area is 100 square degrees. Therefore

$$\text{overall brightness} = \frac{\text{number of visible stars in constellation}}{\text{size of the constellation in square degrees}} \times 100$$

The two numbers in the above equation may be found as part of other lists within this book. Overall brightness for a particular constellation, then, is the average number of visible stars per 100 square degrees.

In many cases, this number makes the constellation no easier to find. Compare two constellations with the same overall brightness, Andromeda and Monoceros. Each constellation has an overall brightness of 7.476, meaning that both have an average of about $7\frac{1}{2}$ visible stars for each 100 square degrees of area. Andromeda is a very easy figure to recognize in the night sky; Monoceros, on the other hand, is not.

Possessive forms of the constellation names (Pages 90–92)

This list gives the possessive case of the names of the constellations. In fact, that is a slight simplification. The list actually gives the Latin 'genitive' case of the constellation names. For ease of understanding, this is often referred to as the possessive case, as that is the most similar English language case comparison available. *Webster's New World Dictionary* defines *genitive* as 'a case in Latin grammar shown by inflection of nouns, pronouns, and adjectives, chiefly expressing possession, origin, source, etc.' With respect to the constellations, this case is most often used to show the location of a star or other object within the boundaries of a certain constellation. Thus, the brightest star in Canis Major, Sirius, is often referred to as alpha Canis Majoris; the tenth variable star discovered in Lyra is referred to as RR Lyrae; and the March, 1992 nova seen within the confines of the constellation of Cygnus is designated Nova Cygni (with or without the year).

Pronunciation key to constellation names (Pages 93–94)

Want to start an argument, or at least a lively discussion? Write the following constellation names on a piece of paper and ask several people to pronounce them: Boötes, Canes Venatici, Equuleus, Gemini, and Ophiuchus. The range of pronunciations you receive will vary greatly.

Such differences go hand in hand with the background and assumptions of the individuals involved. One person may approach the name directly, pronouncing each vowel according to its position within the word, comparing that position to other, more familiar words. Division of syllables may be simple for some, and difficult for others. For example, does 'Canes' have one syllable, or two?

Finally, there is the question of accent. Which syllable will receive the emphasis? In 'Ophiuchus,' is the accent on the syllable 'phi' or is the 'u' stressed?

Such questions have plagued educators and lecturers for centuries. This may be one reason that, in a number of older works, authors refer to the meaning of the constellations, rather than to the names themselves. The problem of pronunciation is eliminated if, when referring to the constellation of Canes Venatici, a writer instead states, 'The unusual star is within the group known as the Hunting Dogs, near the Great Bear.'

Today, professional astronomers are not troubled with such details. When specifying an object, it is sufficient to refer to its right ascension and declination. Still, many of us need to have a standardized listing of constellation pronunciations, which is why this list is included. The primary reference for this list is the report prepared by the Committee of the American Astronomical Society on Preferred Spellings and Pronunciations. It was approved for publication at the New Haven meeting of the A.A.S. in June, 1942.

The pronunciations given are American English.

Proper motions – the 200 stars with the largest proper motion (Pages 95–100)

Proper motion is defined as the rate of change of the apparent position of a star across the celestial sphere. Since everything in the universe is actually in motion, it is impossible to define proper motion as an absolute. Proper motions are usually assigned to very nearby stars. The motions of such stars are measured against very distant stars or other distant objects such as galaxies, whose positional changes are much less apparent. Proper motions are angular measurements and are expressed in terms of arcseconds per year. Since such angles are very small and difficult to measure, it is only after the passage of a number of years (usually, at least 25) that the star's motion becomes apparent and a proper motion is assigned.

The size of the constellations (Pages 101–105)

Every time I scan this list, I find another surprise. I still find it difficult to believe that Sagittarius is almost twice the size of Scorpius. Even Cancer is larger than Scorpius! Aries and Capricornus are about the same size, Canis Major is 50% larger than Ursa Minor, and Orion is only 26th in size. The size of the constellations is given in square degrees, first by rank, and then alphabetically by constellation.

To find the number of square degrees in the entire sky, use the formula for the area of a sphere, $4\pi r^2$, where $r = 1$ radian ($180/\pi$ (or 57.29577951) degrees). Total area of the celestial sphere = 41252.96125 square degrees. This information is provided to demonstrate how the number for '% of sky' was calculated for each constellation.

Solar conjunction dates for the constellations (Pages 106–108)

These two lists give, first by date and then alphabetically by constellation, the approximate date on which the Sun is at the same right ascension as the central point of the constellation. In many cases, this renders the constellation invisible on and around the date listed. The exceptions are those constellations which are circumpolar for your latitude. Thus, Ursa Minor is visible in the northern hemisphere each night of the year, including those nights around its 'conjunction' with the Sun, 21 September. No date is given for Octans, as that constellation completely surrounds the south celestial pole.

Star designations (Pages 109–110)

Modern-day designations (excluding proper names such as 'Arcturus') of the visible stars found in the constellations date from the year 1603. It was in that year that Johannes Bayer published the *Uranometria*, which was essentially an atlas of the constellations. He plotted the positions of just over 2000 stars, including the 1005 stars from the expanded catalog of Tycho Brahe, which was the most accurate catalog of star positions up to that time. Bayer differed from previous methods of stellar nomenclature, however. Until Bayer, stars had been designated by their positions within the mythological figures of the constellations. Thus if a star, for example, Aldebaran, was supposed to represent 'the eye of Taurus the Bull,' that was exactly how it was described.

Bayer's system utilized the letters of the Greek alphabet, in sequence, to differentiate the brightnesses of stars within the same constellation. Thus, the brightest star within a particular constellation was designated by Bayer as α (alpha); second brightest was β (beta); and so on, through to ω (omega). Bayer was not always the most accurate labeler, therefore discrepancies do exist. For example, the α star in Orion, Betelgeuse, is not as bright as the β star, Rigel. Also, on occasion, Bayer would employ a sequential system which was not linked to brightness. An example of this can be seen within the constellation of Ursa Major. The seven prominent stars often referred to as 'The Big Dipper' are labeled by Bayer with the first seven letters of the Greek alphabet: α, β, γ, δ, ε, ξ, and η. When the number of the stars in a constellation exceeded 24 (the number of letters in the Greek alphabet), Bayer used small Roman letters and then capital Roman letters. Since Bayer's map was published, other stellar cartographers have subdivided some of his original designations. For example, in Orion we now find the stars π^1, π^2, π^3, etc. Not withstanding these divisions, it is important to note that no new Greek letters have been added.

**Star names
(Pages 111–125)**

The major question regarding this list was not 'Where to start?' but 'Where to end?' Many listings of star names, old and new, were scanned to obtain this list. In all, 850 star names and variants are provided. Column 1 gives the star's name, with popular variant spellings in parentheses. Note that some stars have more than one name (such as Alpheratz and Sirrah for α And, Alruccabah, Cynosaura, Lodestar, Mismar, Navigatoria, Phoenice, Polaris, Tramontana, and Yilduz for α UMi, etc.). Column 2 provides the star's designation. This is generally the Bayer letter with the constellation name. Column 3 gives the star's visual, or apparent, brightness to the nearest tenth of a magnitude.

It may be noted that certain names are used for different stars (see, for example, Aladfar, Deneb el Okab, Zuben Hakrabi). These are not misprints. There are two reasons for this. In early times, differentiation of star names was not as precise as we might imagine. If two stars were found in the hand of a certain stellar figure, both might have been called 'the hand.' The second reason has a more modern root. It involves the mislabeling of certain stars by stellar cartographers. Then, if that cartographer's maps were used as the basis for others, the mistaken name would be perpetuated, but the original name would also continue on other maps.

**Sun signs – the
constellations of
the zodiac
(Page 126)**

What's your sign? Actually, as these lists demonstrate, there's less than a one-in-eight chance that the Sun was 'in' the constellation given by any astrological horoscope. The reason for this lies in the long-term motion of the Earth called precession. First described by Hipparchus in the second century BC, precession involves a continuous westward motion of the equinoxes along the ecliptic. One complete precessional cycle takes 25 800 years. Thus, Sun sign dates which were established over 2000 years ago are now out of sync by almost one-twelfth the circumference of the sky. This means that the 'signs' are displaced with respect to their dates by almost one constellation, and matters continue to get worse. The next agreement between the sky and the dates given by astrologers will not occur for almost 24 000 years!

Also of interest is the fact that the actual time the sun spends in each constellation varies greatly. Note that the Sun spends 45 days within the boundaries of the constellation of Virgo, 18 days in Ophiuchus and only 7 days in Scorpius. Generally, horoscopes list between 29 and 32 days as the time the Sun is 'in' each sign.

The visibility of the constellations (Pages 127–129)

This list was born out of totally practical reasons. It is often beneficial to know whether or not a particular constellation is visible from a certain location on Earth without having to work out the necessary mathematics.

To use the list, the only quantity required is the latitude of the observer. For any constellation, if the latitude falls within the value given in the first column the entire constellation may be seen from that latitude. If the latitude falls within the range of the numbers given in column 2, the constellation is invisible at that latitude. If the latitude chosen is outside the range of either column, only a portion of the constellation may be seen.

The possibility of visibility is given as a simple parameter, independent of time of night or date. Sidereal time, position of the Sun, etc., must all be taken into account to determine specific visibility on a certain date or at a certain time.

Example(s): From Kansas City (latitude = +39°), what are the possibilities of observing the constellations Capricornus, Centaurus, and Crux at some point during the year?

Capricornus – Since the value for the specified latitude falls within the range of column 1, this entire constellation is visible from Kansas City.

Centaurus – The value for the latitude of Kansas City falls outside the range given in column 1. This means that the entire constellation is not visible from Kansas City. The specified latitude also falls outside the range given in column 2. Therefore, a portion of this constellation is visible from Kansas City.

Crux – The value for latitude falls outside the range of column 1. Again, this implies that the entire constellation is not visible from Kansas City. In addition, the value falls within the range of column 2. Therefore, this constellation is not visible at any time from Kansas City.

One of the interesting sidelights to this list is the number of constellations with the label 'portions visible worldwide.' No fewer than 16 constellations are visible, in part, from *any* location on Earth.

The number of visible stars in the constellations (Pages 130–132)

Column 1 of this list provides the constellation name. Column 2 gives the number of stars in that constellation brighter than visual magnitude 2.4. Column 3 provides the number of stars between visual magnitudes 2.4 and 4.4. Column 4 shows the number of stars between visual magnitudes 4.4 and 5.5. Finally, column 5 totals the numbers in columns 2, 3, and, 4 to give the total number of (normally) visible stars within the constellation.

Alphabetical list of the constellations (with meanings)

Andromeda	The Princess of Ethiopia	Horologium	The Pendulum Clock
Antlia	The Air Pump	Hydra	The Water Snake
Apus	The Bird of Paradise	Hydrus	The Southern Water Snake
Aquarius	The Water Bearer	Indus	The American Indian
Aquila	The Eagle	Lacerta	The Lizard
Ara	The Altar	Leo	The Lion
Aries	The Ram	Leo Minor	The Lion Cub
Auriga	The Charioteer	Lepus	The Hare
Boötes	The Bear Driver	Libra	The Scales
Caelum	The Sculptor's Chisel	Lupus	The Wolf
Camelopardalis	The Giraffe	Lynx	The Lynx
Cancer	The Crab	Lyra	The Harp
Canes Venatici	The Hunting Dogs	Mensa	The Table Mountain
Canis Major	The Greater Dog	Microscopium	The Microscope
Canis Minor	The Lesser Dog	Monoceros	The Unicorn
Capricornus	The Sea Goat	Musca	The Fly
Carina	The Keel (of Argo Navis)	Norma	The Carpenter's Square
Cassiopeia	The Queen of Ethiopia	Octans	The Octant
Centaurus	The Centaur	Ophiuchus	The Serpent Bearer
Cepheus	The King of Ethiopia	Orion	The Hunter
Cetus	The Sea Monster (Whale)	Pavo	The Peacock
Chamaeleon	The Chameleon	Pegasus	The Winged Horse
Circinus	The Compasses	Perseus	Perseus (the hero)
Columba	Noah's Dove	Phoenix	The Phoenix
Coma Berenices	The Hair of Berenice	Pictor	The Painter's Easel
Corona Australis	The Southern Crown	Pisces	The Fishes
Corona Borealis	The Northern Crown	Piscis Austrinus	The Southern Fish
Corvus	The Crow	Puppis	The Stern (of Argo Navis)
Crater	The Cup	Pyxis	The Compass (of Argo Navis)
Crux	The (Southern) Cross		
Cygnus	The Swan	Reticulum	The Net
Delphinus	The Dolphin (Porpoise)	Sagitta	The Arrow
Dorado	The Swordfish	Sagittarius	The Archer
Draco	The Dragon	Scorpius	The Scorpion
Equuleus	The Foal	Sculptor	The Sculptor's Workshop
Eridanus	The River	Scutum	The Shield
Fornax	The Laboratory Furnace	Serpens	The Serpent
Gemini	The Twins	Sextans	The Sextant
Grus	The Crane	Taurus	The Bull
Hercules	Hercules (the hero)	Telescopium	The Telescope

16

Triangulum	The Triangle	Vela	The Sail (of Argo Navis)
Triangulum Australe	The Southern Triangle	Virgo	The Virgin
Tucana	The Toucan	Volans	The Flying Fish
Ursa Major	The Great Bear	Vulpecula	The Fox
Ursa Minor	The Bear Cub		

Abbreviations of constellation names

(Alphabetized by abbreviation, *not* constellation title)

And	Andromeda		Hor	Horologium
Ant	Antlia		Hya	Hydra
Aps	Apus		Hyi	Hydrus
Aql	Aquila		Ind	Indus
Aqr	Aquarius		Lac	Lacerta
Ara	Ara		Leo	Leo
Ari	Aries		Lep	Lepus
Aur	Auriga		Lib	Libra
Boo	Boötes		LMi	Leo Minor
Cae	Caelum		Lup	Lupus
Cam	Camelopardalis		Lyn	Lynx
Cap	Capricornus		Lyr	Lyra
Car	Carina		Men	Mensa
Cas	Cassiopeia		Mic	Microscopium
Cen	Centaurus		Mon	Monoceros
Cep	Cepheus		Mus	Musca
Cet	Cetus		Nor	Norma
Cha	Chamaeleon		Oct	Octans
Cir	Circinus		Oph	Ophiuchus
CMa	Canis Major		Ori	Orion
CMi	Canis Minor		Pav	Pavo
Cnc	Cancer		Peg	Pegasus
Col	Columba		Per	Perseus
Com	Coma Berenices		Phe	Phoenix
CrA	Corona Australis		Pic	Pictor
CrB	Corona Borealis		PsA	Piscis Austrinus
Crt	Crater		Psc	Pisces
Cru	Crux		Pup	Puppis
Crv	Corvus		Pyx	Pyxis
CVn	Canes Venatici		Ret	Reticulum
Cyg	Cygnus		Scl	Sculptor
Del	Delphinus		Sco	Scorpius
Dor	Dorado		Sct	Scutum
Dra	Draco		Sex	Sextans
Equ	Equuleus		Ser	Serpens
Eri	Eridanus		Sge	Sagitta
For	Fornax		Sgr	Sagittarius
Gem	Gemini		Tau	Taurus
Gru	Grus		Tel	Telescopium
Her	Hercules		TrA	Triangulum Australe

Tri	Triangulum		Vel	Vela
Tuc	Tucana		Vir	Virgo
UMa	Ursa Major		Vol	Volans
UMi	Ursa Minor		Vul	Vulpecula

Asterisms within the constellations

1 IN ALPHABETICAL ORDER

The Arc – ε, ζ, and η Ursae Majoris.
The Asses and the Manger – γ and δ Cancri and M44 (the Praesepe).
The Baseball Diamond – β Pegasi, α Andromedae, γ Pegasi, and α Pegasi.
The Belt (of Orion) – δ, ε, and ζ Orionis.
The Bier – α, β, γ, and δ Ursae Majoris.
The Big Dipper – α, β, γ, δ, ε, ζ, and η Ursae Majoris.
The Bull of Poniatowski – 66, 67, 68, and 70 Ophiuchi.
The Butterfly (of Hercules) – β, δ, ε, ζ, η, and π Herculis.
The Butterfly (of Orion) – ζ, ε, δ, γ, α, β, and κ Orionis.
The Circlet (of Pisces) – γ, b, θ, ι, 19, λ, and κ Piscium.
The Diamond (of Virgo) – α Virginis, β Leonis, α Boötis, and α Canum Venaticorum.
The False Cross – ι and ε Carinae and κ and δ Velorum
The Family (of Aquila) – α, β, and γ Aquilae.
The Fish Hook – σ, α, τ, ε, μ^1, ζ$_2$, η, θ, ι^1, κ, λ, and υ Scorpii.
Frederik's Glory – ι, κ, λ, and ψ Andromedae.
The Great Square – α, β, and γ Pegasi and α Andromedae.
The Guardians of the Pole – β and γ Ursae Minoris.
The Head (of Cetus) – α, γ, ξ^2, μ, and λ Ceti.
The Head (of Draco) – β, γ, ξ, and ν Draconis.
The Head (of Hydra) – δ, ε, ζ, η, ρ, and σ Hydrae.
The Heavenly G – α Aurigae, α Geminorum, β Geminorum, α Canis Minoris, α Canis Majoris, β Orionis, α Tauri, and α Orionis.
The Horse and Rider – ζ and 80 Ursae Majoris.
The Hyades – α, γ, δ, and ε Tauri.
The Ice Cream Cone – α, ε, δ, β, γ, and ρ Boötis.
Job's Coffin – α, β, γ, and δ Delphini.
The Keystone – ε, ζ, η, and π Herculis.
The Kids – ε, ζ, and η Aurigae.
The Kite – α, ε, δ, β, γ, and ρ Boötis.
The Large (or, Giant) Dipper – α (or, β) Persei, γ, β, and α Andromedae, γ, α, and β Pegasi.
The Little Dipper – α, δ, ε, ζ, η, γ, and β Ursae Minoris.
The Lozenge – β, γ, ξ, and ν Draconis.
The Milk Dipper – ζ, τ, σ, φ, and λ Sagittarii.
The Northern Cross – α, β, γ, δ, and ε Cygni.
The Northern Fly – 35, 39, and 41 Arietis.
The Pleiades – 17, 19, 20, 23, 27, and η Tauri.
The Pointers – α and β Ursae Majoris.
The Rake – δ, ε, and ζ Orionis.

The Sail – β, δ, γ, and ε Corvi.
The Segment (of Perseus) – η, γ, α, δ, ε, and ζ Persei.
The Sickle – α, η, γ, ζ, μ, and ε Leonis.
The Southern Pointers – α and β Centauri.
The Spring Triangle – α Boötis, α Virginis, and β Leonis.
The Summer Triangle – α Lyrae, α Cygni, and α Aquilae.
The Sword (of Orion) – θ and ι Orionis.
The Three Guides – α Andromedae, β Cassiopeiae, and γ Pegasi.
The Three Kings – δ, ε, and ζ Orionis.
The Three Patriarchs – α, β, and γ Trianguli Australis
The Trapezoid – β, γ, δ, and μ Boötis.
The V (of Taurus) – ε, δ, γ, θ, and α Tauri.
Venus' Mirror – δ, ε, ζ, η, ι, and η Orionis.
The Water Jar – γ, η, π, and ζ Aquarii.
The Winter Octagon – α Aurigae, α Geminorum, β Geminorum, α Canis Minoris, α Canis Majoris, β Orionis, α Tauri, and α Orionis.
The Winter Oval – α Aurigae, α Geminorum, β Geminorum, α Canis Minoris, α Canis Majoris, β Orionis, α Tauri, and α Orionis.
The Winter Triangle – α Canis Majoris, α Canis Minoris, and α Orionis.
The Y (of Virgo) – α, γ, δ, ε, η, and β Virginis.

2 BY CONSTELLATION

Andromeda	The Baseball Diamond (partially)
	Frederik's Glory
	The Great Square (partially)
	The Large Dipper (partially)
	The Three Guides (partially)
Aquarius	The Water Jar
Aquila	The Family
	The Summer Triangle (partially)
Aries	The Northern Fly
Auriga	The Heavenly G (partially)
	The Winter Octagon (partially)
	The Kids
	The Winter Oval (partially)
Boötes	The Diamond (of Virgo) (partially)
	The Ice Cream Cone
	The Kite
	The Spring Triangle (partially)
	The Trapezoid
Cancer	The Asses and the Manger
Canes Venatici	The Diamond (of Virgo) (partially)
Canis Major	The Heavenly G (partially)
	The Winter Octagon (partially)

	The Winter Oval (partially)
	The Winter Triangle (partially)
Canis Minor	The Heavenly G (partially)
	The Winter Octagon (partially)
	The Winter Oval (partially)
	The Winter Triangle (partially)
Carina	The False Cross (partially)
Cassiopeia	The Three Guides (partially)
Centaurus	The Southern Pointers
Cetus	The Head
Corvus	The Sail
Cygnus	The Northern Cross
	The Summer Triangle (partially)
Delphinus	Job's Coffin
Draco	The Head
	The Lozenge
Gemini	The Heavenly G (partially)
	The Winter Octagon (partially)
	The Winter Oval (partially)
Hercules	The Butterfly
	The Keystone
Hydra	The Head
Leo	The Diamond (of Virgo) (partially)
	The Sickle
	The Spring Triangle (partially)
Lyra	The Summer Triangle (partially)
Ophiuchus	The Bull of Poniatowski
Orion	The Belt
	The Butterfly
	The Heavenly G (partially)
	The Rake
	The Sword
	The Three Kings
	Venus' Mirror
	The Winter Oval (partially)
	The Winter Octagon (partially)
	The Winter Triangle (partially)
Pegasus	The Baseball Diamond (partially)
	The Great Square (partially)
	The Large Dipper (partially)
	The Three Guides (partially)
Perseus	The Large Dipper (partially)
	The Segment
Pisces	The Circlet
Sagittarius	The Milk Dipper
Scorpius	The Fish Hook
Taurus	The Heavenly G (partially)

	The Hyades
	The Pleiades
	The V
	The Winter Octagon (partially)
	The Winter Oval (partially)
Triangulum Australe	The Three Patriarchs
Ursa Major	The Arc
	The Bier
	The Big Dipper
	The Horse and Rider
	The Pointers
Ursa Minor	The Guardians of the Pole
	The Little Dipper
Vela	The False Cross (partially)
Virgo	The Diamond (partially)
	The Spring Triangle (partially)
	The Y

Biblical references to the constellations and stars

In the entire Bible, there are only three specific references to Constellations: (quotes are from the New American Standard version)

Job 9:7–9
Who commands the sun not to shine,
And sets a seal upon the stars;
Who alone stretches out the heavens,
And tramples down the waves of the sea;
Who makes the Bear, Orion, and the Pleiades,
And the chambers of the south.

Job 38:31–3
Can you bind the chains of the Pleiades,
Or loose the cords of Orion?
Can you lead forth a constellation in its season,
And guide the Bear with her satellites?
Do you know the ordinances of the heavens,
Or fix their rule over the earth?

Amos 5:8
He who made the Pleiades and Orion
And changes deep darkness into morning,
Who also darkens day into night,
Who calls for the waters of the sea
And pours them out on the surface of the earth,
The Lord is His name.

Other *astronomical* references:

In the Old Testament

Gen. 1: 1	Gen. 1: 14–19	Gen. 2: 1
Gen. 2: 4	Exod. 20: 11	Exod. 31: 17
Deut. 4: 19	Deut. 17: 3	Neh. 9: 6
Job 22: 12	Job 26: 7	Job 37: 18
Job 38: 1–7	Job 38: 13	Ps. 8: 3–5
Ps. 19: 1–6	Ps. 65: 8	Ps. 72: 5
Ps. 74: 16–17	Ps. 89: 11	Ps. 89: 36–7
Ps. 102: 25–6	Ps. 104: 2	Ps. 104: 5
Ps. 104: 19	Ps. 136: 7–9	Ps. 147: 4
Ps. 148: 1–6	Prov. 3: 19	Prov. 8: 27–31
Is. 13: 10	Is. 13: 13	Is. 24: 23
Is. 30: 26	Is. 34: 4	Is. 37: 16
Is. 40: 12	Is. 40: 22	Is. 40: 26
Is. 42: 5	Is. 44: 24	Is. 45: 12
Is. 45: 18	Is. 47: 13	Is. 48: 13
Is. 51: 6	Is. 51: 13	Is. 60: 19–20
Jer. 10: 2	Jer. 10: 12	Jer. 27: 5
Jer. 31: 35	Jer. 33: 20–21	Jer. 33: 22
Jer. 33: 25–6	Jer. 51: 15	Ezek. 32: 7–8
Joel 2: 10	Joel 2: 31	Joel 3: 15
Amos 8: 9	Mic. 3: 6	

In the New Testament

Matt. 2: 2	Matt. 2: 7	Matt. 2: 9–10
Matt. 24: 29	Matt. 24: 35	Mark 13: 24–5
Luke 21: 25	Acts 2: 19–20	Acts 17: 24
1 Cor. 15: 40–41	Heb. 1: 10–12	2 Pet. 1: 19
2 Pet. 3: 8	2 Pet. 3: 10	Jude 1: 13
Rev. 1: 16	Rev. 6: 12–14	Rev. 8: 10–12
Rev. 21: 23	Rev. 22: 5	

Bordering constellations

Andromeda: Cassiopeia (N), Lacerta (W), Pegasus (S & W), Perseus (N & E), Pisces (S & E), Triangulum (S)

Antlia: Centaurus (E), Hydra (N & E), Pyxis (W), Vela (S)

Apus: Ara (N), Chamaeleon (W), Circinus (N), Musca (W), Octans (S & E), Pavo (E), Triangulum Australe (N)

Aquarius: Aquila (W), Capricornus (S & W), Cetus (E), Delphinus (N), Equuleus (N), Pegasus (N), Pisces (N & E), Piscis Austrinus (S), Sculptor (S)

Aquila: Aquarius (E), Capricornus (S & E), Delphinus (N & E), Hercules (N & W), Ophiuchus (W), Sagitta (N), Sagittarius (S), Scutum (S & W), Serpens (S & W)

Ara: Apus (S), Corona Australis (N), Norma (W), Pavo (E), Scorpius (N), Telescopium (E), Triangulum Australe (S & W)

Aries: Cetus (S), Perseus (N), Pisces (W), Taurus (E), Triangulum (N)

Auriga: Camelopardalis (N), Gemini (S & E), Lynx (N & E), Perseus (W), Taurus (S)

Boötes: Canes Venatici (W), Coma Berenices (W), Corona Borealis (E & S), Draco (N), Hercules (E), Serpens (E), Ursa Major (N & W), Virgo (S & W)

Caelum: Columba (E), Dorado (S), Eridanus (N & W), Horologium (W), Lepus (N), Pictor (S & E)

Camelopardalis: Auriga (S), Cassiopeia (W), Cepheus (N & W), Draco (S & E), Lynx (S & E), Perseus (S), Ursa Major (S & E), Ursa Minor (N & E)

Cancer: Canis Minor (S & W), Gemini (W), Hydra (S), Leo (E), Lynx (N)

Canes Venatici: Boötes (E), Coma Berenices (S), Ursa Major (N, E, W)

Canis Major: Columba (S & W), Lepus (W), Monoceros (N), Puppis (S & E)

Canis Minor: Cancer (N & E), Gemini (N), Hydra (E), Monoceros (S & W)

Capricornus: Aquarius (N & E), Aquila (N & W), Microscopium (S), Piscis Austrinus (S), Sagittarius (W)

Carina: Centaurus (N & E), Chamaeleon (S), Musca (E), Pictor (S & W), Puppis (N), Vela (N & E), Volans (S & W)

Cassiopeia: Andromeda (S), Camelopardalis (E), Cepheus (N & W), Lacerta (W), Perseus (S & E)

Centaurus: Antlia (W), Carina (S & W), Circinus (S & E), Crux (S, E, W), Hydra (N), Lupus (S & E), Musca (S), Vela (W)

Cepheus: Camelopardalis (S & E), Cassiopeia (S & E), Cygnus (S & W), Draco (W), Lacerta (S), Ursa Minor (N & W)

Cetus: Aquarius (W), Aries (N), Eridanus (S & E), Fornax (S), Pisces (N & W), Sculptor (S), Taurus (E)

Chamaeleon: Apus (E), Carina (N), Mensa (W), Musca (N), Octans (S), Volans (N)

Circinus: Apus (S), Centaurus (N & W), Lupus (N), Musca (W), Norma (N & E), Triangulum Australe (E)

Columba: Caelum (W), Canis Major (N & E), Lepus (N), Pictor (S), Puppis (S & E)

Coma Berenices: Boötes (E), Canes Venatici (N), Leo (W), Ursa Major (N & W), Virgo (S)

Corona Australis: Ara (S), Sagittarius (N & E), Scorpius (W), Telescopium (S)

Corona Borealis: Boötes (W & N), Hercules (N & E), Serpens (S)

Corvus: Crater (W), Hydra (S), Virgo (N & E)

Crater: Corvus (E), Hydra (S & W), Leo (N), Sextans (W), Virgo (N & E)

Crux: Centaurus (N, E, W), Musca (S)

Cygnus: Cepheus (N & E), Draco (N & W), Lacerta (E), Lyra (W), Pegasus (S & E), Vulpecula (S)

Delphinus: Aquarius (S), Aquila (S & W), Equuleus (S & E), Pegasus (E), Sagitta (N & W), Vulpecula (N)

Dorado: Caelum (N), Horologium (N & W), Hydrus (W), Mensa (S), Pictor (N & E), Reticulum (S & W), Volans (E)

Draco: Bootes (S), Camelopardalis (N & W), Cepheus (E), Cygnus (S & E), Hercules (S), Lyra (S), Ursa Major (S & W), Ursa Minor (N, E, W)

Equuleus: Aquarius (S), Delphinus (N & W), Pegasus (N & E)

Eridanus: Caelum (S & E), Cetus (N & W), Fornax (S, W, N), Horologium (S & E), Hydrus (S), Lepus (E), Orion (N & E), Phoenix (N & W), Taurus (N)

Fornax: Cetus (N), Eridanus (N, E, S), Phoenix (S), Sculptor (W)

Gemini: Auriga (N & W), Cancer (E), Canis Minor (S), Lynx (N), Monoceros (S), Orion (S & W), Taurus (W)

Grus: Indus (S & W), Microscopium (W), Phoenix (E), Piscis Austrinus (N), Sculptor (N & E), Tucana (S)

Hercules: Aquila (S & E), Boötes (W), Corona Borealis (S & W), Draco (N), Lyra (N & E), Ophiuchus (S & E), Sagitta (E), Serpens (S & E), Vulpecula (E)

Horologium: Caelum (E), Dorado (S & E), Eridanus (N & W), Hydrus (S & W), Reticulum (S & E)

Hydra: Antlia (S & W), Cancer (N), Canis Minor (W), Centaurus (S), Corvus (N), Crater (N & E), Leo (N), Libra (N & E), Monoceros (W), Puppis (S & W), Pyxis (S & W), Sextans (E & N), Virgo (N)

Hydrus: Dorado (E), Eridanus (N), Horologium (N & E), Mensa (S & E), Octans (S & W), Reticulum (N), Tucana (N & W)

Indus: Grus (N & E), Microscopium (N), Octans (S), Pavo (S & W), Telescopium (W), Tucana (N & E)

Lacerta: Andromeda (E), Cassiopeia (E), Cepheus (N), Cygnus (W), Pegasus (S)

Leo: Cancer (W), Coma Berenices (E), Crater (S), Hydra (S), Leo Minor (N), Sextans (S & W), Ursa Major (N), Virgo (S & E)

Leo Minor: Leo (S), Lynx (N & W), Ursa Major (N & E)

Lepus: Caelum (S), Canis Major (E), Columba (S), Eridanus (W), Monoceros (N), Orion (N)

Libra: Hydra (S & W), Lupus (S), Ophiuchus (E), Scorpius (S & E), Serpens (N), Virgo (N & W)

Lupus: Centaurus (N & W), Circinus (S), Libra (N), Norma (S & E), Scorpius (N & E)

Lynx: Auriga (W & S), Camelopardalis (N & W), Cancer (S), Gemini (S), Leo Minor (E), Ursa Major (E & N)

Lyra: Cygnus (E), Draco (N), Hercules (S & W), Vulpecula (S & E)

Mensa: Chamaeleon (E), Dorado (N), Hydrus (N & W), Octans (S), Volans (N & E)

Microscopium: Capricornus (N), Grus (E), Indus (S), Piscis Austrinus (E), Sagittarius (W)

Monoceros: Canis Major (S), Canis Minor (N & E), Gemini (N), Hydra (E), Lepus (S), Orion (N & W), Puppis (S)

Musca: Apus (E), Carina (W), Centaurus (N), Chamaeleon (S), Circinus (E), Crux (N)

Norma: Ara (E), Circinus (S & W), Lupus (N & W), Scorpius (N & E), Triangulum Australe (S)

Octans: Apus (N & W), Chamaeleon (N), Hydrus (N & E), Indus (N), Mensa (N), Pavo (N), Tucana (N)

Ophiuchus: Aquila (E), Hercules (N & W), Libra (W), Sagittarius (E), Scorpius (S & W), Serpens (N, E, S, W)

Orion: Eridanus (S & W), Gemini (N & E), Lepus (S), Monoceros (S & E), Taurus (N & W)

Pavo: Apus (W), Ara (N & W), Indus (N & E), Octans (S), Telescopium (N)

Pegasus: Andromeda (N & E), Aquarius (S), Cygnus (N & W), Delphinus (W), Equuleus (S & W), Lacerta (N), Pisces (S & E), Vulpecula (N & W)

Perseus: Andromeda (S & W), Aries (S), Auriga (E), Camelopardalis (N), Cassiopeia (N & W), Taurus (S), Triangulum (S & W)

Phoenix: Eridanus (S & E), Fornax (N), Grus (W), Sculptor (N), Tucana (S)

Pictor: Caelum (N & W), Carina (N & E), Columba (N), Dorado (S & W), Puppis (E), Volans (S)

Pisces: Andromeda (N & W), Aquarius (S & W), Aries (E), Cetus (S & E), Pegasus (N & W), Triangulum (N & E)

Piscis Austrinus: Aquarius (N), Capricornus (N), Grus (S), Microscopium (W), Sculptor (N & E)

Puppis: Canis Major (N & W), Carina (S), Columba (N & W), Hydra (N & E), Monoceros (N), Pictor (W), Pyxis (E), Vela (S & E)

Pyxis: Antlia (E), Hydra (N & E), Puppis (W), Vela (S)

Reticulum: Dorado (N & E), Horologium (N & W), Hydrus (S)

Sagitta: Aquila (S), Delphinus (S & E), Hercules (W), Vulpecula (N)

Sagittarius: Aquila (N), Capricornus (E), Corona Australis (S & W), Microscopium (E), Ophiuchus (W), Scorpius (S & W), Scutum (N & W), Serpens (N), Telescopium (S)

Scorpius: Ara (S), Corona Australis (E), Libra (N & W), Lupus (S & W), Norma (S & W), Ophiuchus (N & E), Sagittarius (N & E)

Sculptor: Aquarius (N), Cetus (N), Fornax (E), Grus (S & W), Phoenix (S), Piscis Austrinus (S & W)

Scutum: Aquila (N & E), Sagittarius (S & E), Serpens (N & W)

Serpens: Aquila (N & E), Boötes (W), Corona Borealis (N), Hercules (N & E), Libra (S), Ophiuchus (N, E, S, W), Sagittarius (S), Scutum (S & E), Virgo (W)

Sextans: Crater (E), Hydra (S & W), Leo (N & E)

Taurus: Aries (W), Auriga (N), Cetus (W), Eridanus (S), Gemini (E), Orion (S & E), Perseus (N)

Telescopium: Ara (W), Corona Australis (N), Indus (E), Pavo (S), Sagittarius (N)

Triangulum: Andromeda (N), Aries (S), Perseus (N & E), Pisces (S & W)

Triangulum Australe: Apus (S), Ara (N & E), Circinus (W), Norma (N)

Tucana: Grus (N), Hydrus (S & E), Indus (S & W), Octans (S), Phoenix (N)

Ursa Major: Boötes (S & E), Camelopardalis (N & W), Canes Venatici (S, E, W), Coma Berenices (S & E), Draco (N & E), Leo (S), Leo Minor (S & W), Lynx (S & W)

Ursa Minor: Camelopardalis (S & W), Cepheus (S & E), Draco (S, E, W)

Vela: Antlia (N), Carina (S & W), Centaurus (E), Puppis (N & W), Pyxis (N)

Virgo: Boötes (N & E), Coma Berenices (N), Corvus (S & W), Crater (S & W), Hydra (S), Leo (N & W), Libra (S & E), Serpens (E)

Volans: Carina (N & E), Chamaeleon (S), Dorado (W), Mensa (S & W), Pictor (N)

Vulpecula: Cygnus (N), Delphinus (S), Hercules (W), Lyra (N & W), Pegasus (S & E), Sagitta (S)

The 200 brightest stars

Rank	Designation	Name	Bright star number	Apparent magnitude	Absolute magnitude	Distance (light yrs)
	—	The Sun	—	−26.7	4.9	0.000016
1	α CMa	Sirius	2491	−1.46	0.7	8.7
2	α Car	Canopus	2326	−0.72	−5.5	300
3	α Cen	Rigil Kentaurus	5459/60	−0.27	4.6	4.4
4	α Boo	Arcturus	5340	−0.04	−0.3	36
5	α Lyr	Vega	7001	0.03	0.3	26
6	α Aur	Capella	1708	0.08	0.1	45
7	β Ori	Rigel	1713	0.12	−7.0	850
8	α CMi	Procyon	2943	0.38	2.8	11
9	α Eri	Achernar	472	0.46	−1.3	75
10	α Ori	Betelgeuse	2061	0.50	−5.5	650
11	β Cen	Hadar	5267	0.61	−4.3	300
12	α Aql	Altair	7557	0.77	2.1	17
13	α Cru	Acrux	4730/1	0.79	−3.8	270
14	α Tau	Aldebaran	1457	0.85	−0.2	65
15	α Sco	Antares	6134	0.96	−4.5	400
16	α Vir	Spica	5056	0.98	−3.2	220
17	β Gem	Pollux	2990	1.14	0.7	35
18	α PsA	Fomalhaut	8728	1.16	1.8	22
19	β Cru	Mimosa	4853	1.25	−4.0	370
20	α Cyg	Deneb	7924	1.25	−7.0	1500
21	α Leo	Regulus	3982	1.35	−1.0	85
22	ε CMa	Adhara	2618	1.50	−5.0	620
23	α Gem	Castor	2890	1.58	0.9	45
24	γ Cru	Gacrux	4763	1.63	−2.5	220
25	λ Sco	Shaula	6527	1.63	−3.3	300
26	γ Ori	Bellatrix	1790	1.64	−4.1	450
27	β Tau	El Nath	1791	1.65	−3.0	270
28	β Car	Miaplacidus	3685	1.68	−0.4	85
29	ε Ori	Alnilam	1903	1.70	−6.8	1600
30	α Gru	Alnair	8425	1.74	0.2	65
31	ζ Ori	Alnitak	1948/9	1.76	−6.6	1600
32	ε UMa	Alioth	4905	1.77	0.2	70
33	α Per	Mirfak	1017	1.79	−4.4	570
34	α UMa	Dubhe	4301	1.79	−0.7	105
35	δ CMa	Wezen	2693	1.84	−7.0	2100
36	ε Sgr	Kaus Australis	6879	1.85	−1.1	125
37	ε Car	Avior	3307	1.86	−3.1	340
38	η UMa	Alkaid	5191	1.86	−2.1	210

Rank	Designation	Name	Bright star number	Apparent magnitude	Absolute magnitude	Distance (light yrs)
39	θ Sco	Sargas	6553	1.87	−1.0	121
40	β Aur	Menkalinan	2088	1.90	−0.3	90
41	γ Leo	Algeiba	4057/8	1.90	−1.4	148
42	α TrA	Atria	6217	1.92	−0.1	80
43	γ Gem	Alhena	2421	1.93	−0.7	105
44	α Pav	Peacock	7790	1.94	−2.8	310
45	β CMa	Mirzam	2294	1.98	−4.8	750
46	α Hya	Alphard	3748	1.98	−0.3	95
47	α Ari	Hamal	617	2.00	0.2	75
48	α UMi	Polaris	424	2.02	−3.4	360
49	σ Sgr	Nunki	7121	2.02	−2.5	250
50	β Cet	Deneb Kaitos	188	2.04	0.8	60
51	α And	Alpheratz	15	2.06	−0.7	120
52	β And	Mirach	337	2.06	0.5	67
53	κ Ori	Saiph	2004	2.06	−2.1	217
54	θ Cen	Menkent	5288	2.06	1.1	50
55	β UMi	Kochab	5563	2.08	−0.5	100
56	α Oph	Rasalhague	6556	2.08	1.2	49
57	β Gru	—	8636	2.10	−3.4	408
58	β Per	Algol	936	2.12	0.4	72
59	β Leo	Denebola	4534	2.14	1.7	39
60	γ Cen	—	4819	2.17	−1.8	204
61	γ Cyg	Sadr	7796	2.20	−5.4	1087
62	λ Vel	Alsuhail	3634	2.21	−1.1	148
63	δ Ori	Mintaka	1852	2.23	−2.0	233
64	α Cas	Schedar	168	2.23	−1.8	204
65	α CrB	Alphecca	5793	2.23	0.5	72
66	γ Dra	Eltanin	6705	2.23	−0.8	130
67	ζ Pup	Naos	3165	2.25	−7.1	2400
68	ι Car	Tureis	3699	2.25	−1.6	192
69	γ¹ And	Alamak	603	2.26	−2.2	251
70	β Cas	Caph	21	2.27	1.6	45
71	ζ¹ UMa	Mizar	5054	2.27	0.6	69
72	ε Sco	—	6241	2.29	−1.0	148
73	ε Cen	—	5132	2.30	−3.9	570
74	α Lup	—	5469	2.30	−3.3	430
75	η Cen	—	5440	2.31	−3.0	390
76	δ Sco	Dschubba	5953	2.32	−4.0	590
77	β UMa	Merak	4295	2.37	1.0	61
78	ε Boo	Pulcherrima	5505/6	2.37	−1.6	204
79	α Phe	Ankaa	99	2.39	0.4	83
80	ε Peg	Enif	8308	2.39	−3.7	543
81	κ Sco	—	6580	2.41	−3.4	470

Rank	Designation	Name	Bright star number	Apparent magnitude	Absolute magnitude	Distance (light yrs)
82	β Peg	Scheat	8775	2.42	−0.9	148
83	η Oph	Sabik	6378	2.43	1.0	63
84	γ UMa	Phecda	4554	2.44	−0.3	116
85	α Cep	Alderamin	8162	2.44	1.6	48
86	η CMa	Aludra	2827	2.45	−7.0	2700
87	ε Cyg	Gienah	7949	2.46	1.2	57
88	γ Cas	Cih	264	2.47	−1.5	204
89	α Peg	Markab	8781	2.49	0.4	86
90	α Vel	—	3734	2.50	−1.9	250
91	α Cet	Menkar	911	2.53	−2.7	362
92	ζ Cen	—	5231	2.55	−3.4	520
93	β¹ Sco	Acrab	5984	2.55	−2.7	362
94	δ Leo	Zosma	4357	2.56	1.0	68
95	ζ Oph	—	6175	2.56	−5.1	1087
96	α Lep	Arneb	1865	2.58	−3.2	466
97	γ Crv	—	4662	2.59	−2.0	300
98	δ Cen	—	4621	2.60	−0.3	125
99	ζ Sgr	Ascella	7194	2.60	−0.4	130
100	β Lib	Zubeneschamali	5685	2.61	−5.0	1080
101	β Ari	Sheratan	553	2.64	2.0	44
102	α Col	Phakt	1956	2.64	−0.2	120
103	β Crv	Kraz	4786	2.65	0.3	96
104	α Ser	Unuk al Hai	5854	2.65	1.3	61
105	δ Cas	Ruchbah	403	2.68	0.5	88
106	η Boo	Mufrid	5235	2.68	2.9	32
107	β Lup	—	5571	2.68	−3.4	540
108	ι Aur	Hasseleh	1577	2.69	−0.7	155
109	μ Vel	—	4216	2.69	−0.6	148
110	α Mus	—	4798	2.69	−2.9	430
111	υ Sco	Lesath	6508	2.69	−3.4	540
112	π Pup	—	2773	2.70	0.2	102
113	δ Sgr	Kaus Meridionalis	6859	2.70	1.1	69
114	γ Aql	Tarazed	7525	2.72	−1.3	204
115	δ Oph	Yed Prior	6056	2.74	0.4	96
116	η Dra	—	6132	2.74	1.3	64
117	γ Vir	Porrima	4825/6	2.75	2.7	33
118	ι Cen	—	5028	2.75	1.7	52
119	α² Lib	Zubenelgenubi	5531	2.75	1.6	56
120	θ Car	—	4199	2.76	−3.9	700
121	ι Ori	Hatsya	1899	2.77	−0.2	130
122	β Oph	Cheleb	6603	2.77	0.4	99
123	γ Lup	—	5776	2.78	−2.7	408
124	β Eri	Cursa	1666	2.79	1.3	65

Rank	Designation	Name	Bright star number	Apparent magnitude	Absolute magnitude	Distance (light yrs)
125	β Dra	Alwaid	6536	2.79	−1.6	250
126	β Hyi	—	98	2.80	3.8	20
127	δ Cru	—	4656	2.80	−4.8	1087
128	ρ Pup	—	3185	2.81	0.5	93
129	ζ Her	—	6212	2.81	2.9	31
130	λ Sgr	Kaus Borealis	6913	2.81	1.4	61
131	τ Sco	Al Niyat	6165	2.82	−0.7	163
132	γ Peg	Algenib	39	2.83	−5.7	1630
133	ε Vir	Vindemiatrix	4932	2.83	1.0	76
134	β Lep	Nihal	1829	2.84	−0.7	163
135	ζ Per	—	1203	2.85	−2.2	326
136	β TrA	—	5897	2.85	2.4	39
137	β Ara	—	6461	2.85	0.5	96
138	α Hyi	—	591	2.86	1.3	68
139	α Tuc	—	8502	2.86	0.1	125
140	η Tau	—	1165	2.87	−2.6	408
141	δ Cyg	—	7528	2.87	0.3	109
142	δ Cap	Deneb Algedi	8322	2.87	2.6	37
143	μ Gem	Tejat Posterior	2286	2.88	−0.6	163
144	ε Per	—	1220	2.89	−2.3	362
145	γ TrA	—	5671	2.89	−2.1	326
146	π Sco	—	5944	2.89	−2.1	326
147	σ Sco	Al Niyat	6084	2.89	−4.4	900
148	π Sgr	Albaldah	7264	2.89	0.0	125
149	β CMi	Gomeisa	2845	2.90	−0.7	171
150	α² CVn	Cor Caroli	4915	2.90	0.1	121
151	θ Eri	Acamar	897/8	2.91	0.6	93
152	β Aqr	Sadalsuud	8232	2.91	−3.2	543
153	γ Per	—	915	2.93	−1.1	204
154	τ Pup	—	2553	2.93	0.0	125
155	η Peg	Matar	8650	2.94	−1.0	192
156	γ Eri	Zaurak	1231	2.95	−2.1	326
157	δ Crv	Algorab	4757	2.95	−0.2	136
158	α Ara	—	6510	2.95	−2.8	466
159	υ Car	—	3890	2.96	0.1	121
160	α Aqr	Sadalmelek	8414	2.96	−1.6	272
161	ε Gem	Mebsuta	2473	2.98	−0.9	192
162	ε Leo	Ras Elased Australis	3873	2.98	−2.0	326
163	ε Aur	Almaaz	1605	2.99	−2.8	466
164	γ² Sgr	Al Nasl	6746	2.99	0.0	130
165	ζ Aql	Deneb el Okab	7235	2.99	1.3	72
166	β Tri	—	622	3.00	−0.3	148
167	ζ Tau	Alcyone	1910	3.00	−2.5	408

Rank	Designation	Name	Bright star number	Apparent magnitude	Absolute magnitude	Distance (light yrs)
168	ε Crv	Minkar	4630	3.00	0.2	121
169	γ Hya	—	5020	3.00	0.2	121
170	δ Per	—	1122	3.01	−1.0	204
171	φ UMa	—	4335	3.01	0.0	130
172	γ Gru	Al Dhanab	8353	3.01	−1.4	250
173	ζ CMa	Furud	2282	3.02	−4.0	815
174	o² CMa	—	2653	3.02	−7.1	3400
175	γ Boo	Seginus	5435	3.03	0.0	130
176	ι¹ Sco	—	6615	3.03	−0.6	171
177	μ Cen	—	5193	3.04	−2.7	470
178	μ UMa	Tania Australis	4069	3.05	0.8	93
179	β Mus	—	4844	3.05	−1.1	217
180	γ UMi	Pherkad	5735	3.05	−4.6	1087
181	δ Dra	Aldib	7310	3.07	0.6	102
182	μ¹ Sco	—	6247	3.08	−2.9	520
183	α¹ Her	Ras Algethi	6406/7	3.08	−5.4	1630
184	β¹ Cyg	Albireo	7417	3.08	−0.8	192
185	β Cap	Dabih	7776	3.08	−1.9	326
186	ζ Hya	—	3547	3.11	0.8	93
187	ν Hya	—	4232	3.11	0.4	116
188	η Sgr	—	6832	3.11	1.4	72
189	α Ind	—	7869	3.11	1.4	71
190	β Col	Wasn	2040	3.12	0.4	116
191	α Lyn	—	3705	3.13	0.1	130
192	N Vel	—	3803	3.13	−0.2	148
193	λ Cen	—	4467	3.13	−2.1	370
194	κ Cen	—	5576	3.13	−2.6	470
195	ζ Ara	—	6285	3.13	1.4	74
196	ι UMa	Talitha	3569	3.14	2.5	43
197	δ Her	Sarin	6410	3.14	1.4	74
198	π Her	—	6418	3.16	0.2	130
199	η Aur	Hoedus II	1641	3.17	−0.1	148
200	h UMa	—	3775	3.17	0.2	48

Central points of the constellations

	Approximate		Exact	
	RA	Dec.	RA	Dec.
Andromeda	1h	+40°	0h46m	+37°
Antlia	10h	−30°	10h14m	−32°
Apus	16h	−75°	16h01m	−75°
Aquarius	22h	−10°	22h15m	−11°
Aquila	20h	+5°	19h37m	+3.5°
Ara	17h	−55°	17h18m	−56.5°
Aries	3h	+20°	2h35m	+20.5°
Auriga	6h	+40°	6h01m	+42°
Boötes	15h	+30°	14h40m	+31°
Caelum	5h	−40°	4h40m	−38°
Camelopardalis	9h	+70°	8h48m	+69°
Cancer	9h	+20°	8h36m	+20°
Canes Venatici	13h	+40°	13h04m	+40.5°
Canis Major	7h	−20°	6h47m	−22°
Canis Minor	8h	+5°	7h36m	+6.5°
Capricornus	21h	−20°	21h00m	−18°
Carina	9h	−65°	8h40m	−63°
Cassiopeia	1h	+60°	1h16m	+62°
Centaurus	13h	−50°	13h01m	−47.5°
Cepheus	2h	+70°	2h15m	+70°
Cetus	2h	−10°	1h38m	−7.5°
Chamaeleon	11h	−80°	10h40m	−79°
Circinus	15h	−60°	14h30m	−62°
Columba	6h	−35°	5h45m	−35°
Coma Berenices	13h	+25°	12h45m	+23.5°
Corona Australis	19h	−40°	18h35m	−41.5°
Corona Borealis	16h	+35°	15h48m	+33°
Corvus	12h	−20°	12h24m	−18°
Crater	11h	−15°	11h21m	−15.5°
Crux	12h	−60°	12h24m	−60°
Cygnus	21h	+45°	20h34m	+44.5°
Delphinus	21h	+10°	20h39m	+11.5°
Dorado	5h	−60°	5h14m	−59.5°
Draco	15h	+65°	15h09m	+67°
Equuleus	21h	+10°	21h08m	+7.5°
Eridanus	3h	−30°	3h15m	−29°
Fornax	3h	−30°	2h46m	−32°
Gemini	7h	+25°	7h01m	+22.5°

	Approximate		Exact	
	RA	Dec.	RA	Dec.
Grus	22h	−45°	22h25m	−47°
Hercules	17h	+30°	17h21m	+27.5°
Horologium	3h	−55°	3h15m	−53.5°
Hydra	11h	−15°	11h33m	−14°
Hydrus	2h	−70°	2h16m	−70°
Indus	22h	−60°	21h55m	−60°
Lacerta	22h	+45°	22h25m	+46°
Leo	11h	+15°	10h37m	+13.5°
Leo Minor	10h	+35°	10h11m	+32.5°
Lepus	6h	−20°	5h31m	−19°
Libra	15h	−15°	15h08m	−15°
Lupus	15h	−45°	15h09m	−42.5°
Lynx	8h	+50°	7h56m	+47.5°
Lyra	19h	+35°	18h49m	+36.5°
Mensa	5h	−80°	5h28m	−77.5°
Microscopium	21h	−35°	20h55m	−36.5°
Monoceros	7h	+00°	7h01m	+0.5°
Musca	13h	−70°	12h31m	−69.5°
Norma	16h	−50°	15h58m	−51°
Octans	—	−85°	—	−82.5°
Ophiuchus	17h	−10°	17h20m	−8°
Orion	6h	+5°	5h32m	+6°
Pavo	20h	−65°	19h33m	−66°
Pegasus	23h	+20°	22h39m	+19°
Perseus	3h	+45°	3h06m	+45°
Phoenix	1h	−50°	0h54m	−49°
Pictor	6h	−55°	5h41m	−53.5°
Pisces	0h	+15°	0h26m	+13°
Piscis Austrinus	22h	−30°	22h14m	−31°
Puppis	7h	−30°	7h14m	−31°
Pyxis	9h	−30°	8h56m	−27°
Reticulum	4h	−60°	3h54m	−60°
Sagitta	20h	+20°	19h37m	+18.5°
Sagittarius	19h	−30°	19h03m	−28.5°
Scorpius	17h	−25°	16h49m	−27°
Sculptor	0h	−35°	0h24m	−32.5°
Scutum	19h	−10°	18h37m	−10°
Serpens	17h	+5°	16h55m	+5°
Sextans	10h	00°	10h14m	−2°
Taurus	5h	+15°	4h39m	+15.5°
Telescopium	19h	−50°	19h16m	−51°
Triangulum	2h	+30°	2h08m	+31°

	Approximate		Exact	
	RA	Dec.	RA	Dec.
Triangulum Australe	16h	−65°	15h59m	−65°
Tucana	0h	−65°	23h43m	−66.5°
Ursa Major	11h	+50°	11h16m	+51°
Ursa Minor	—	+80°	—	+77.5°
Vela	10h	−45°	9h43m	−47°
Virgo	13h	−5°	13h21m	−4°
Volans	8h	−70°	7h48m	−69.5°
Vulpecula	20h	+25°	20h12m	+24°

Directional extremes

	North	South	West	East
Andromeda	+53°	+21°	22h56m	2h36m
Antlia	−24°	−40°	9h25m	11h03m
Apus	−67°	−83°	13h45m	18h17m
Aquarius	+3°	−25°	20h36m	23h54m
Aquila	+19°	−12°	18h38m	20h36m
Ara	−45°	−68°	16h31m	18h06m
Aries	+31°	+10°	1h44m	3h27m
Auriga	+56°	+28°	4h35m	7h27m
Boötes	+55°	+7°	13h33m	15h47m
Caelum	−27°	−49°	4h18m	5h03m
Camelopardalis	+85°	+53°	3h11m	14h25m
Cancer	+33°	+7°	7h53m	9h19m
Canes Venatici	+53°	+28°	12h04m	14h05m
Canis Major	−11°	−33°	6h09m	7h26m
Canis Minor	+13°	00°	7h04m	8h09m
Capricornus	−8°	−28°	20h04m	21h57m
Carina	−51°	−75°	6h02m	11h18m
Cassiopeia	+78°	+46°	22h56m	3h36m
Centaurus	−30°	−65°	11h03m	14h59m
Cepheus	+89°	+51°	20h01m	8h30m
Cetus	+10°	−25°	23h55m	3h21m
Chamaeleon	−75°	−83°	7h32m	13h48m
Circinus	−54°	−70°	13h35m	15h26m
Columba	−27°	−43°	5h03m	6h28m
Coma Berenices	+34°	+13°	11h57m	13h33m
Corona Australis	−37°	−46°	17h55m	19h15m
Corona Boraelis	+40°	+26°	15h14m	16h22m
Corvus	−11°	−25°	11h54m	12h54m
Crater	−6°	−25°	10h48m	11h54m
Crux	−55°	−65°	11h53m	12h55m
Cygnus	+61°	+28°	19h07m	22h01m
Delphinus	+21°	+2°	20h13m	21h06m
Dorado	−49°	−70°	3h52m	6h36m
Draco	+86°	+48°	9h18m	21h00m
Equuleus	+13°	+2°	20h54m	21h23m
Eridanus	00°	−58°	1h22m	5h09m
Fornax	−24°	−40°	1h44m	3h48m
Gemini	+35°	+10°	5h57m	8h06m
Grus	−37°	−57°	21h25m	23h25m
Hercules	+51°	+4°	15h47m	18h56m

	North	South	West	East
Horologium	−40°	−67°	2h12m	4h18m
Hydra	+7°	−35°	8h08m	14h58m
Hydrus	−58°	−82°	0h02m	4h33m
Indus	−45°	−75°	20h25m	23h25m
Lacerta	+57°	+35°	21h55m	22h56m
Leo	+33°	−6°	9h18m	11h56m
Leo Minor	+42°	+23°	9h19m	11h04m
Lepus	−11°	−27°	4h54m	6h09m
Libra	00°	−30°	14h18m	15h59m
Lupus	−30°	−55°	14h13m	16h05m
Lynx	+62°	+33°	6h13m	9h40m
Lyra	+48°	+25°	18h12m	19h26m
Mensa	−70°	−85°	3h20m	7h37m
Microscopium	−28°	−45°	20h25m	21h25m
Monoceros	+12°	−11°	5h54m	8h08m
Musca	−64°	−75°	11h17m	13h46m
Norma	−42°	−60°	15h25m	16h31m
Octans	−75°	−90°	0h	24h
Ophiuchus	+14°	−30°	15h58m	18h42m
Orion	+23°	−11°	4h41m	6h23m
Pavo	−57°	−75°	17h37m	21h30m
Pegasus	+36°	+2°	21h06m	0h13m
Perseus	+59°	+31°	1h26m	4h46m
Phoenix	−40°	−58°	23h24m	2h24m
Pictor	−43°	−64°	4h32m	6h51m
Pisces	+33°	−7°	22h49m	2h04m
Piscis Austrinus	−25°	−37°	21h25m	23h04m
Puppis	−11°	−51°	6h02m	8h26m
Pyxis	−17°	−37°	8h26m	9h26m
Reticulum	−53°	−67°	3h14m	4h35m
Sagitta	+21°	+16°	18h56m	20h18m
Sagittarius	−12°	−45°	17h41m	20h25m
Scorpius	−8°	−46°	15h44m	17h55m
Sculptor	−25°	−40°	23h04m	1h44m
Scutum	−4°	−16°	18h18m	18h56m
Serpens	+26°	−16°	14h55m	18h56m
Sextans	+7°	−11°	9h39m	10h49m
Taurus	+31°	00°	3h20m	5h58m
Telescopium	−45°	−57°	18h06m	20h26m
Triangulum	+37°	+25°	1h29m	2h48m
Triangulum Australe	−60°	−70°	14h50m	17h09m
Tucana	−57°	−76°	22h05m	1h22m
Ursa Major	+73°	+29°	8h05m	14h27m
Ursa Minor	+90°	+65°	0h	24h

	North	South	West	East
Vela	−37°	−57°	8h02m	11h24m
Virgo	+14°	−22°	11h35m	15h08m
Volans	−64°	−75°	6h35m	9h02m
Vulpecula	+29°	+19°	18h56m	21h28m

Extinct constellations

Antinoüs   **Original composition and location:** η, σ, θ, ι, κ, λ, ν, and δ Aql.

Figure represented: Servant (homosexual slave and lover) of the Roman Emperor Hadrian, Antinoüs drowned himself in the Nile, believing that, in this sacrifice, his master's life might be prolonged. Formed by Gerard Mercator on his globe of 1551. According to a ninth century Hyginus–Cicero manuscript (and others who published later maps), the figure represents Ganymede, brought to his lover, Jove, by Aquila.

Apes   **Original composition and location:** Flamsteed 41, 33, 35, 39, in Aries.

Figure represented: A fly. (This has sometimes been referred to as Musca.)

Apis   **Original composition and location:** Occupied the position now held by the constellation Musca.

Figure represented: A bee.

Apparatus Chemicus   See Fornax Chemica.

Argo Navis   **Original composition and location:** A huge constellation which has been subdivided into the three present-day constellations Carina, Puppis, and Vela. This was one of the 48 original constellations described by Ptolemy. It is the only one he described which is no longer used.

Figure represented: The ship Argo, built for Jason, which carried him and his crew in search of the golden fleece. When the voyage was over, Athena, who had been the crew's special protectoress, placed the ship in the sky.

The Battery of Volta   **Original composition and location:** Two 4th-magnitude stars located between the heads of Delphinus, Equuleus, and Pegasus. Thomas Young, British physician, physicist, and egyptologist, invented this constellation in 1806, to honor the important invention by Alessandro Volta.

Figure represented: Volta's trough battery.

Cancer Minor   **Original composition and location:** Formed by Plancius, around 1614, of stars between Cancer and Gemini.

Figure represented: A small crab.

Cesaries Another name for the constellation Coma Berenices; sometimes referred to as Cesaries Berenices.

Cerberus **Original composition and location:** Flamsteed 93, 95, 96, and 109, lying halfway between the head of Hercules and the head of Cygnus. Formed by Hevelius in 1687.

Figure represented: Cerberus, the three-headed hound that guarded the gates of Hades.

Cor Caroli **Original composition and location:** Occupied the position of the present-day constellation of Canes Venatici. This name was subsequently given to the brightest star of that constellation (α CVn).

Figure represented: The (crowned) heart of Charles I. The invention of this figure is credited to Sir Charles Scarborough, physician of Charles I of England. It first appeared on the planisphere of Francis Lamb in 1673. It was first published (being called Cor Caroli Regis Martyris) in 1675 by the British poet Edward Sherburne.

Corona Firmiana Vulgo Septentrionalis **Original composition and location:** Occupied the same position as Corona Borealis.

Figure represented: In 1730, Corbinianus Thomas, German Benedictine monk and professor of mathematics and theology, replaced Corona Boraelis with this constellation to honor the Archbishop of Salzburg, Leopold Anton von Firmian.

Custos Messium **Original composition and location:** Formed from some inconspicuous stars not far from the pole, between Camelopardalis, Cassiopeia, and Cepheus.

Figure represented: The harvest-keeper. Possibly refers to a play on words by Lalande, originator of this constellation, to honor his friend, Charles Messier. Lalande introduced this constellation on his celestial globe of 1779.

Felis **Original composition and location:** Formed from stars between Antlia and Hydra.

Figure represented: A cat. This constellation was suggested by Lalande, and first appeared in the 1801 atlas of J. E. Bode.

Fornax Chemica Original name of the present-day constellation Fornax. Introduced by Lacaille in 1754–5.

Friedrichs Ehre (Frederici Honores) **Original composition and location:** Formed in 1787 out of the present stars of Lacerta and a few from Andromeda by Bode, commemorating Frederick II of Prussia.

43

Figure represented: A sword, pen, and olive branch, wreathed in a laurel.

Gallus **Original composition and location:** Formed by Plancius in 1613 out of stars between Argo Navis and Canis Major.

Figure represented: The cock which crowed twice after Peter denied Jesus three times.

Gladii Electorales Saxonici **Original composition and location:** Bounded by α Boo, α Ser, β Lib, μ Vir and τ Vir.

Figure represented: The crossed swords of the Electors of Saxony. Introduced in 1688 by the German astronomer Godfried Kirch, honoring the Emperor of Germany, Leopold I.

Globus Aerostaticus **Original composition and location:** Proposed by Lalande and introduced by Bode in 1801 to honor the Montgolfier brothers, pioneers in ballooning. Formed out of stars in the southern part of Capricornus.

Figure represented: A hot air balloon.

Horologium Oscillatorium Former name for the present constellation Horologium.

Jordanus (or, Iordanus) **Original composition and location:** Formed by Plancius in 1613 of stars to the east, south, and west of Ursa Major.

Figure represented: The Jordan River.

Leo Palatinus **Original composition and location:** Formed by Karl-Joseph König, astronomer of the observatory at Mannheim, Germany from 1782 to 1786 from faint stars between the present-day constellations of Aquarius and Aquila.

Figure represented: An imperial lion, honoring König's patrons, Charles Theodore and his wife, Elisabeth Augusta. Introduced in 1785.

Lilium (Fleur de Lis) **Original composition and location:** Formed out of four stars north of Aries, which Plancius had grouped into the constellation Apes.

Figure represented: A flower. Known in the Middle Ages as the 'Flower of the Virgin Mary.' It was also a sign for the Holy Trinity of Christendom. The design of this flower was often used as a symbol of loyalty on shields, tapestries, amulets, etc.

Lochium Funis **Original composition and location:** Formed by Bode in 1801, near Argo Navis out of some stars in the present-day constellation of Pyxis.

Figure represented: A nautical log line.

Machina Electrica   **Original composition and location:** South of the central portion of Cetus.

Figure represented: In 1801, Bode placed an electrostatic generator here.

Marmor Sculptile   **Original composition and location:** Formed in 1810 by William Croswell out of the stars of the present-day constellation of Reticulum.

Figure represented: The bust of Christopher Columbus.

Mons Maenalus   **Original composition and location:** At the feet of Boötes. On the 1687 map of Hevelius (on which it was introduced), Boötes is standing on this constellation.

Figure represented: the mountain.

Musca Australis   This title was substituted by Lacaille, about 1752, for Apis (see Apis). It is now known as the constellation of Musca.

Musca Boraelis   **Original composition and location:** Flamsteed 41, 33, 35, 39, in Aries.

Figure represented: A fly. Introduced by Plancius around 1614 under the name of Apes (see Apes).

Noctua   **Original composition and location:** On the extreme tail-tip of Hydra.

Figure represented: Elijah Hinsdale Burritt, in his star atlas of 1833, represents this constellation as a night owl. It occupied the same position as Turdus Solitarius (see Turdus Solitarius).

Officina Typographica   **Original composition and location:** Formed by Bode in 1801 from stars directly east of Sirius. Honored the 350th anniversary of the invention of movable type.

Figure represented: A printing office.

Phoenicopterus   The flamingo. Alternative seventeenth century name for the present constellation Grus.

Polophylax   **Original composition and location:** Formed by Plancius, around 1614, from stars between Crux and Piscis Austrinus.

Figure represented: Apparently, Plancius used a combination of two Greek words in his creation of this constellation: polos (πολοσ), a pivot on which anything turns, apparently meant to represent the axis of the celestial sphere, and phylax (φυλαξ or φυλασσω) , watcher, guard, or sentinel. This figure, then was the 'guardian of the pole,' an apt description for a constellation positioned so near the south celestial pole.

Pomum Imperiale

Original composition and location: Composed of seven faint stars to the southeast of the head of Aquila.

Figure represented: The orb of Emperor Leopold I of Germany. Invented in 1688 by Godfried Kirch, German astronomer.

Psalterium Georgianum

Original composition and location: Formed in 1789 by the Jesuit astronomer Maximilian Hell, it lies between the forefeet of Taurus and Eridanus. Its brightest star is o^2 Eri.

Figure represented: The harp of King George III of England. Hell published his maps to honor Sir William Herschel (and his recent discovery of Uranus) and Herschel's patron, George III.

Quadrans Muralis

Original composition and location: Between the right foot of Hercules, the left hand of Boötes, and Draco.

Figure represented: The mural quadrant of Lalande, which he used to chart 50 000 stars while at the College de France. Introduced in 1795 by J. Fortin, maker of globes and spheres for the French royal family.

Renne (le Renne)

Original composition and location: Formed of faint stars between Cassiopeia and Camelopardalis. Some star maps have this figure labeled 'Tarandus vel Rangifer.'

Figure represented: A reindeer. Pierre-Charles le Monnier invented this constellation in 1743. It commemorated an expedition in which he participated to measure the length of a degree of the Earth's latitude near the north pole.

Robur Carolinum

Original composition and location: 25 stars, including β Carinae.

Figure represented: Formed by Halley in 1679 to represent the Royal Oak of King Charles II of England, in which the king had lain hidden for 24 hours after his defeat by Cromwell on the battle of Worcester.

Sagitta Australe

Original composition and location: Formed by Plancius, around 1614, out of stars north of Aquila.

Figure represented: An arrow.

Sceptrum

Original composition and location: A few faint stars between the forelegs of Pegasus and the head of Cepheus.

Figure represented: The French scepter and hand of justice. In 1679, Royer attempted to replace the already established (and still-existing) constellation of Lacerta with this figure honoring Louis XIV.

Sceptrum Brandenburgicum

Original composition and location: Five stars, between 4th and 5th magnitude, in a straight line north and south, below the first bend in Eridanus, west of Lepus, southwest of Rigel.

Figure represented: The Brandenburg Scepter. Introduced in 1688 by the German astronomer Godfried Kirch.

Sciurus Volans

Original composition and location: Two stars in the tail of the present-day constellation of Camelopardalis.

Figure represented: The flying squirrel. Introduced by William Croswell of Boston in 1810.

Solarium

Original composition and location: East of Horologium, between the head of Hydrus and the tail of Dorado.

Figure represented: A sundial.

Solitaire

Original composition and location: Formed of 41 stars between 3rd and 9th magnitude by Le Monnier in 1776, near the tip of the tail of Hydra. Also known as Turdus Solitarius, this group was later changed by some to Noctua, the Night Owl (see Noctua). In his initial map, Le Monnier gave the boundaries of this constellation as long. $= 0° - 26°30'$ ($0°$ defined as beginning at the western boundary of Scorpius) and lat. $=$ ecliptic $- 15°$S.

Figure represented: a Solitary Thrush.

Tarandus vel Rangifer

See Renne.

Taurus Poniatowski

Original composition and location: Formed out of 16 faint stars of Ophiuchus, near the borders of Aquila and Hercules, in 1777, by the Abbé Poczobut, of Wilna, Poland.

Figure represented: The bull of Stanislaus Poniatowski (Stanislaus II), King of Poland.

Tigris

Original composition and location: Formed by Plancius, around 1614, of stars between Cygnus and Aquila.

Figure represented: The Tigris River.

Triangulum Minor   **Original composition and location:** Formed of three faint stars, just to the south of Triangulum, toward α Ari.

Figure represented: A small triangle. Included by Hevelius in his atlas of 1687.

Tubus Herschelii Major   **Original composition and location:** π Gem, ψ Aur, and stars near Lynx.

Figure represented: Formed in 1789 by Maximilian Hell, Jesuit astronomer, in honor of Sir William Herschel's discovery of Uranus. Hell placed two now-extinct constellations, (the other being Tubus Herschelii Minor), on either side of the point in the sky (in Gemini) where Herschel first discovered the planet Uranus in 1781. This constellation represented Herschel's 20-foot reflecting telescope.

Tubus Herschelii Minor   **Original composition and location:** Formed from faint stars just to the east of the Hyades in Taurus.

Figure represented: Father Hell, in 1789, used these stars to represent the seven-foot reflecting telescope of Sir William Herschel. (See also Tubus Herschelii Major.)

Turdus Solitarius   See Solitaire.

Vulpecula cum Ansere   Formed by Hevelius to represent a fox and a goose, later simplified to Vulpecula.

The magnitude system

Magnitude Difference	Brightness ratio
0.1	1.0964782
0.2	1.2022644
0.25	1.2589254
0.3	1.3182567
0.333	1.3593563
0.4	1.4454397
0.5	1.5848932
0.6	1.7378008
0.666	1.8478497
0.7	1.9054607
0.75	1.9952623
0.8	2.0892961
0.9	2.2908677
1.0	**2.5118865**
1.5	3.9810719
2.0	6.3095738
2.5	**10.000000**
3.0	15.848932
3.5	25.118865
4.0	39.810719
4.5	63.095738
5.0	**100.00000**
5.5	158.48932
6.0	251.18865
6.5	398.10719
7.0	630.95738
7.5	**1000.0000**
8.0	1584.8932
8.5	2511.8865
9.0	3981.0719
9.5	6309.5738
10.0	**10 000.000**
11.0	25 118.865
12.0	63 095.738
12.5	**100 000**
13.0	158 489.32
14.0	398 107.19
15.0	**1 000 000.0**
16.0	2 511 886.5
17.0	6 309 573.8

17.5	**10 000 000**
18.0	15 848 932
19.0	39 810 719
20.0	**100 000 000**

Messier objects

1 IN NUMERICAL ORDER

Number	Popular name	Constellation	Magnitude	Type of object	NGC number
M1	CrabNebula	Tau	8.4	Supernova remnant	1952
M2		Aqr	6.4	Globular cluster	7089
M3		CVn	6.3	Globular cluster	5272
M4		Sco	6.5	Globular cluster	6121
M5		Ser	6.1	Globular cluster	5904
M6		Sco	5.5	Galactic cluster	6405
M7		Sco	3.3	Galactic cluster	6475
M8	Lagoon Nebula	Sgr	5.1	Diffuse nebula	6523
M9		Oph	8.0	Globular cluster	6333
M10		Oph	6.7	Globular cluster	6254
M11	Wild Duck	Sct	6.8	Galactic cluster	6705
M12		Oph	6.6	Globular cluster	6218
M13	Great Cluster	Her	5.9	Globular cluster	6205
M14		Oph	8.0	Globular cluster	6402
M15		Peg	6.4	Globular cluster	7078
M16	Star Queen or Eagle Nebula	Ser	6.6	Galactic cluster	6611
M17	Omega or Swan Nebula	Sgr	7.5	Galactic cluster	6618
M18		Sgr	7.2	Galactic cluster	6613
M19		Oph	6.9	Globular cluster	6273
M20	Trifid Nebula	Sgr	8.5	Diffuse nebula	6514
M21		Sgr	6.5	Galactic cluster	6531
M22		Sgr	5.6	Globular cluster	6656
M23		Sgr	5.9	Galactic cluster	6494
M24		Sgr	4.6	Galactic cluster	6603
M25		Sgr	6.2	Galactic cluster	—
M26		Sct	9.3	Galactic cluster	6694
M27	Dumbbell Nebula	Vul	7.6	Planetary nebula	6853
M28		Sgr	7.6	Globular cluster	6626
M29		Cyg	8.0	Galactic cluster	6913
M30		Cap	7.7	Globular cluster	7099
M31	Great Galaxy in Andromeda	And	3.5	Spiral galaxy	224
M32		And	8.2	Elliptical galaxy	221
M33	Pinwheel Galaxy	Tri	5.8	Spiral galaxy	598
M34		Per	5.8	Galactic cluster	1039
M35		Gem	5.6	Galactic cluster	2168

Number	Popular name	Constellation	Magnitude	Type of object	NGC number
M36		Aur	6.5	Galactic cluster	1960
M37		Aur	6.2	Galactic cluster	2099
M38		Aur	7.0	Galactic cluster	1912
M39		Cyg	5.3	Galactic cluster	7092
M40	double star	UMa	8.9	Double star	—
M41		CMa	5.0	Galactic cluster	2287
M42	Orion Nebula	Ori	4.0	Diffuse nebula	1976
M43		Ori	9.0	Diffuse nebula	1982
M44	Praesepe or Beehive	Cnc	3.9	Galactic cluster	2632
M45	Pleiades or Seven Sisters	Tau	1.6	Galactic cluster	—
M46		Pup	6.6	Galactic cluster	2437
M47		Pup	5.0	Galactic cluster	2422
M48		Hya	6.0	Galactic cluster	2548
M49		Vir	8.5	Elliptical galaxy	4472
M50		Mon	6.3	Galactic cluster	2323
M51	Whirlpool Galaxy	CVn	8.4	Spiral galaxy	5194
M52		Cas	8.2	Galactic cluster	7654
M53		Com	7.8	Globular cluster	5024
M54		Sgr	7.8	Globular cluster	6715
M55		Sgr	6.2	Globular cluster	6809
M56		Lyr	8.7	Globular cluster	6779
M57	Ring Nebula	Lyr	9.0	Planetary nebula	6720
M58		Vir	9.9	Spiral galaxy	4579
M59		Vir	10.0	Spiral galaxy	4621
M60		Vir	9.0	Elliptical galaxy	4649
M61		Vir	9.6	Spiral galaxy	4303
M62		Oph	6.6	Globular cluster	6266
M63		CVn	8.9	Spiral galaxy	5055
M64	Black Eye Galaxy	Com	8.5	Spiral galaxy	4826
M65		Leo	9.4	Spiral galaxy	3623
M66		Leo	9.0	Spiral galaxy	3627
M67		Cnc	6.1	Galactic cluster	2682
M68		Hya	8.2	Globular cluster	4590
M69		Sgr	8.0	Globular cluster	6637
M70		Sgr	8.1	Globular cluster	6681
M71		Sge	7.6	Globular cluster	6838
M72		Aqr	9.3	Globular cluster	6981
M73		Aqr	9.1	Galactic cluster	6994
M74		Psc	9.3	Spiral galaxy	628
M75		Sgr	8.6	Globular cluster	6864
M76	Little Dumbbell or Cork Nebula	Per	11.4	Planetary Nebula	650/ 651
M77		Cet	8.9	Spiral galaxy	1068

Number	Popular name	Constellation	Magnitude	Type of object	NGC number
M78		Ori	8.3	Diffuse nebula	2068
M79		Lep	7.5	Globular cluster	1904
M80		Sco	7.5	Globular cluster	6093
M81		UMa	7.0	Spiral galaxy	3031
M82	Exploding Galaxy	UMa	8.4	Irregular galaxy	3034
M83		Hya	7.6	Spiral galaxy	5236
M84		Vir	9.4	Elliptical galaxy	4374
M85		Com	9.3	Elliptical galaxy	4382
M86		Vir	9.2	Elliptical galaxy	4406
M87		Vir	8.7	Elliptical galaxy	4486
M88		Com	9.5	Spiral galaxy	4501
M89		Vir	10.3	Elliptical galaxy	4552
M90		Vir	9.6	Spiral galaxy	4569
M91		Com	9.5	Spiral galaxy	4548
M92		Her	6.4	Globular cluster	6341
M93		Pup	6.5	Galactic cluster	2447
M94		CVn	8.3	Spiral galaxy	4736
M95		Leo	9.8	Spiral galaxy	3351
M96		Leo	9.3	Spiral galaxy	3368
M97	Owl Nebula	UMa	11.1	Planetary nebula	3587
M98		Com	10.2	Spiral galaxy	4192
M99		Com	9.9	Spiral galaxy	4254
M100		Com	9.4	Spiral galaxy	4321
M101		UMa	7.9	Spiral galaxy	5457
M102	Duplicate observation of M101?	UMa (?)	—	—	
M103		Cas	6.9	Galactic cluster	581
M104	Sombrero Galaxy	Vir	8.3	Spiral galaxy	4594
M105		Leo	9.7	Elliptical galaxy	3379
M106		CVn	8.4	Spiral galaxy	4258
M107		Oph	9.2	Globular cluster	6171
M108		UMa	10.5	Spiral galaxy	3556
M109		UMa	10.0	Spiral galaxy	3992
M110		And	9.4	Elliptical galaxy	205

2 BY CONSTELLATION

Andromeda:	M31, M32, M110
Aquarius:	M2, M72, M73
Auriga:	M36, M37, M38
Cancer:	M44, M67
Canes Venatici:	M3, M51, M63, M94, M106
Canis Major:	M41
Capricornus:	M30
Cassiopeia:	M52, M103
Cetus:	M77
Coma Berenices:	M53, M64, M85, M88, M91, M98, M99, M100
Cygnus:	M29, M39
Gemini:	M35
Hercules:	M13, M92
Hydra:	M48, M68, M83
Leo:	M65, M66, M95, M96, M105
Lepus:	M79
Lyra:	M56, M57
Monoceros:	M50
Ophiuchus:	M9, M10, M12, M14, M19, M62, M107
Orion:	M42, M43, M78
Pegasus:	M15
Perseus:	M34, M76
Pisces:	M74
Puppis:	M46, M47, M93
Sagitta:	M71
Sagittarius:	M8, M17, M18, M20, M21, M22, M23, M24, M25, M28, M54, M55, M69, M70, M75
Scorpius:	M4, M6, M7, M80
Scutum:	M11, M26
Serpens:	M5, M16
Taurus:	M1, M45
Triangulum:	M33
Ursa Major:	M40, M81, M82, M97, M101, M108, M109
Virgo:	M49, M58, M59, M60, M61, M84, M86, M87, M89, M90, M104
Vulpecula:	M27

3 BY RIGHT ASCENSION

M-object	NGC number	Right ascension	Declination
M110	205	00h38m	+41°25′
M32	221	00h40m	+40°36′
M31	224	00h40m	+41°00′
M103	581	01h30m	+60°27′
M33	598	01h31m	+30°24′
M74	628	01h34m	+15°32′
M76	650/651	01h39m	+51°19′
M34	1039	02h39m	+42°34′
M77	1068	02h40m	−00°13′
M45	—	03h44m	+23°58′
M79	1904	05h22m	−24°34′
M38	1912	05h25m	+35°48′
M1	1952	05h32m	+21°59′
M36	1960	05h32m	+34°07′
M42	1976	05h33m	−05°21′
M43	1982	05h33m	−05°18′
M78	2068	05h44m	+00°12′
M37	2099	05h49m	+32°33′
M35	2168	06h05m	+24°20′
M41	2287	06h45m	−20°42′
M50	2323	07h01m	−08°16′
M47	2422	07h34m	−14°22′
M46	2437	07h40m	−14°42′
M93	2447	07h42m	−23°45′
M48	2548	08h11m	−05°38′
M44	2632	08h38m	+19°52′
M67	2682	08h48m	+12°00′
M81	3031	09h52m	+69°18′
M82	3034	09h52m	+69°56′
M95	3351	10h41m	+11°58′
M96	3368	10h44m	+12°05′
M105	3379	10h45m	+12°51′
M108	3556	11h09m	+55°57′
M97	3587	11h12m	+55°18′
M65	3623	11h16m	+13°23′
M66	3627	11h18m	+13°17′
M109	3992	11h55m	+53°39′
M98	4192	12h11m	+15°11′
M99	4254	12h16m	+14°42′
M106	4258	12h17m	+47°35′
M61	4303	12h19m	+04°45′
M40	—	12h20m	+58°22′
M100	4321	12h20m	+16°06′

M-object	NGC number	Right ascension	Declination
M84	4374	12h22m	+13°10′
M85	4382	12h23m	+18°28′
M86	4406	12h24m	+13°13′
M49	4472	12h27m	+08°16′
M87	4486	12h28m	+12°40′
M88	4501	12h30m	+14°42′
M91	4548	12h33m	+14°46′
M89	4552	12h33m	+12°50′
M90	4569	12h34m	+13°26′
M58	4579	12h35m	+12°05′
M68	4590	12h37m	−26°29′
M104	4594	12h37m	−11°21′
M59	4621	12h40m	+11°55′
M60	4649	12h41m	+11°49′
M94	4736	12h49m	+41°23′
M64	4826	12h54m	+21°57′
M53	5024	13h10m	+18°26′
M63	5055	13h14m	+42°17′
M51	5194	13h28m	+47°27′
M83	5236	13h34m	−29°37′
M3	5272	13h40m	+28°38′
M101	5457	14h01m	+54°35′
M102	5866	15h05m	+55°57′
M5	5904	15h16m	+02°16′
M80	6093	16h14m	−22°52′
M4	6121	16h21m	−26°24′
M107	6171	16h30m	−12°57′
M13	6205	16h40m	+36°33′
M12	6218	16h45m	−01°52′
M10	6254	16h55m	−04°02′
M62	6266	16h58m	−30°03′
M19	6273	17h00m	−26°11′
M9	6333	17h16m	−18°28′
M92	6341	17h16m	+43°12′
M14	6402	17h35m	−03°13′
M6	6405	17h37m	−32°11′
M7	6475	17h51m	−34°48′
M23	6494	17h54m	−19°01′
M20	6514	17h59m	−23°02′
M8	6523	18h02m	−24°20′
M21	6531	18h02m	−22°30′
M24	6603	18h16m	−18°27′
M16	6611	18h16m	−13°48′
M18	6613	18h17m	−17°09′

M-object	NGC number	Right ascension	Declination
M17	6618	18h18m	−16°12′
M28	6626	18h22m	−24°54′
M69	6637	18h28m	−32°23′
M25	—	18h29m	−19°17′
M22	6656	18h33m	−23°58′
M70	6681	18h40m	−32°21′
M26	6694	18h43m	−09°27′
M11	6705	18h48m	−06°20′
M54	6715	18h52m	−30°32′
M57	6720	18h52m	+32°58′
M56	6779	19h15m	+30°05′
M55	6809	19h37m	−31°03′
M71	6838	19h52m	+18°39′
M27	6853	19h57m	+22°35′
M75	6864	20h03m	−22°04′
M29	6913	20h22m	+38°21′
M72	6981	20h51m	−12°44′
M73	6994	20h56m	−12°50′
M15	7078	21h28m	+11°57′
M39	7092	21h30m	+48°13′
M2	7089	21h31m	−01°03′
M30	7099	21h38m	−23°25′
M52	7654	23h22m	+61°20′

Meteor showers

1 BY DATE

Approximate date of maximum	Shower	Radiant		ZHR
3 Jan	Quadrantids	15h21m	+48.5°	80
10 Jan	Coma Berenicids	11h40m	+25°	8
16 Jan	δ Cancrids	08h24m	+20°	7
8 Feb	α Centaurids	14h00m	−59°	10
26 Feb	δ Leonids	10h36m	+19°	24
16 Mar	Corona Australids	18 h19m	−42°	8
26 Mar	Virginids	12h24m	0°	6
9 Apr	α Virginids	13h16m	−13°	8
17 Apr	σ Leonids	13h00m	−5°	12
22 Apr	April Lyrids[a]	18h06m	+33.6°	12
23 Apr	π Puppids	07h20m	−45°	10
25 Apr	μ Virginids	14h44m	−5°	7
28 Apr	α Boötids	14h32m	+19°	8
1 May	φ Boötids	16h00m	+51°	6
3 May	α Scorpiids	16h00m	−22°	6
3 May	η Aquarids	22h22m	−1.9°	60
3 Jun	τ Herculids	15h12m	+39°	15
5 Jun	X Scorpiids	16h28m	−13°	10
7 Jun	Daytime Arietids	02h56m	+23°	50
7 Jun	Daytime ζ Perseids	04h08m	+23°	40
8 Jun	Librids[b]	15h09m	−28.3°	10
11 Jun	Sagittariids[c]	20h16m	−35°	30
13 Jun	θ Ophiuchids	17h48m	−28°	2
16 Jun	June Lyrids[d]	18h32m	+35°	9
26 Jun	Corvids[e]	12h48m	−19.1°	13
28 Jun	Draconids	16 h 55m	+56°	5
28 Jun	June Boötids	14h36m	+49°	6
29 Jun	Daytime β Taurids	05h44m	+19°	25
9 Jul	ε Pegasids	22h40m	+15°	8
14 Jul	July Phoenicids[f]	02h05m	−47.9°	30
16 Jul	o Draconids	18h04m	+59°	3
22 Jul	Capricornids	20h52m	−23°	4
29 Jul	S. δ Aquarids	22h12m	−16.5°	30
30 Jul	α Capricornids[g]	20h28m	−10°	30
5 Aug	S. ι Aquarids	22h13m	−14.7°	15
12 Aug	Perseids	03h05m	+57.4°	95
12 Aug	N. δ Aquarids	22h36m	−5°	20

18 Aug	κ Cygnids	19h04m	+59°	5
1 Sep	Aurigids	05h39m	+42°	30
20 Sep	N. ι Aquarids	21h48m	−6°	15
20 Sep	S. Piscids	00h24m	0°	10
21 Sep	κ Aquarids	22h32m	−5°	5
29 Sep	Daytime Sextantids	10h08m	0°	30
3 Oct	Annual Andromedids	00h20m	+8°	13
	Annual Andromedids	01h20m	+34°	10
9 Oct	October Draconids	17h28m	+54.1°	2
12 Oct	N. Piscids	01h44m	+14°	6
19 Oct	ε Geminids	06h56m	+27°	5
21 Oct	Orionids	06h18m	+15.8°	30
24 Oct	Leo Minorids	10h48m	+37°	3
3 Nov	S. Taurids[h]	03h22m	+13.6°	7
12 Nov	Pegasids	22h20m	+21°	5
13 Nov	N. Taurids[h]	03h53m	+22.3°	7
17 Nov	Leonids	10h09m	+22.2°	15
27 Nov	Andromedids[i]	01h40m	+44°	—
5 Dec	December Phoenicids	01h00m	−55°	100
	December Phoenicids[j]	01h00m	−45°	100
10 Dec	Monocerotids	06h39m	+14°	3
10 Dec	S. × Orionids	05h40m	+16°	8
11 Dec	N. × Orionids	05h36m	+26°	4
11 Dec	σ Hydrids	08h26m	+1.6°	5
11 Dec	δ Arietids	03h28m	+22°	5
14 Dec	Geminids	07h29m	+32.5°	90
22 Dec	Ursids	14h28m	+75.85°	20

[a] Generally very weak, at the threshold of visual detection.
[b] Appeared only in 1937.
[c] Appeared in 1958.
[d] This shower has appeared only from 1966 onward.
[e] Appeared only in 1937.
[f] Observed only by radar from 1953–8.
[g] Not resolvable visually from the Southern δ Aquarids.
[h] These two showers cannot be resolved from one another visually.
[i] This shower varies wildly. The best showing was 27 Nov 1885, with a ZHR of 13 000.
[j] Appeared only in 1965.

2 BY CONSTELLATION

Constellation	Meteor shower	Approximate date of maximum
Andromeda	Annual Andromedids	3 Oct
	Andromedids	27 Nov
Aquarius	η Aquarids	3 May
	S. δ Aquarids	29 Jul
	S. ι Aquarids	5 Aug
	N. δ Aquarids	12 Aug
	N. ι Aquarids	20 Sep
	κ Aquarids	21 Sep
Aries	Daytime Arietids	7 Jun
	δ Arietids	11 Dec
Auriga	Aurigids	1 Sep
Boötes	Quadrantids	3 Jan
	α Boötids	28 Apr
	φ Boötids	1 May
	June Boötids	28 Jun
Cancer	δ Cancrids	16 Jan
Capricornus	Capricornids	22 Jul
	α Capricornids	30 Jul
Centaurus	η Centaurids	8 Feb
Coma Berenices	Coma Berenicids	10 Jan
Corona Australis	Corona Australids	16 Mar
Corvus	Corvids	26 Jun
Cygnus	κ Cygnids	18 Aug
Draco	Draconids	28 Jun
	o Draconids	16 Jul
	October Draconids	9 Oct
Gemini	ε Geminids	19 Oct
	Geminids	14 Dec
Hercules	τ Herculids	3 Jun
Hydra	σ Hydrids	11 Dec
Leo	δ Leonids	26 Feb
	σ Leonids	17 Apr
	Leonids	17 Nov
Leo Minor	Leo Minorids	24 Oct
Libra	Librids	8 Jun
Lyra	April Lyrids	22 Apr
	June Lyrids	16 Jun
Monoceros	Monocerotids	10 Dec
Ophiuchus	θ Ophiuchids	13 Jun
Orion	Orionids	21 Oct
	S. χ Orionids	10 Dec
	N. χ Orionids	11 Dec

Constellation	Meteor shower	Approximate date of maximum
Pegasus	ε Pegasids	9 Jul
	Pegasids	12 Nov
Perseus	Daytime ζ Perseids	7 Jun
	Perseids	12 Aug
Phoenix	July Phoenicids	14 Jul
	December Phoenicids	5 Dec
Pisces	S. Piscids	20 Sep
	N. Piscids	12 Oct
Puppis	η Puppids	23 Apr
Sagittarius	Sagittariids	11 Jun
Scorpius	α Scorpiids	3 May
	χ Scorpiids	5 Jun
Sextans	Daytime Sextantids	29 Sep
Taurus	Daytime β Taurids	29 Jun
	S. Taurids	3 Nov
	N. Taurids	13 Nov
Ursa Major	Ursids	22 Dec
Virgo	Virginids	26 Mar
	α Virginids	9 Apr
	μ Virginids	25 Apr

Midnight culmination dates of the constellations

1 BY MONTH

January:

2 Jan	Canis Major
5 Jan	Gemini
5 Jan	Monoceros
8 Jan	Puppis
14 Jan	Canis Minor
18 Jan	Volans
19 Jan	Lynx
30 Jan	Cancer
31 Jan	Carina

February:

4 Feb	Pyxis
13 Feb	Vela
22 Feb	Sextans
23 Feb	Leo Minor
24 Feb	Antlia

March:

1 Mar	Chamaeleon
1 Mar	Leo
11 Mar	Ursa Major
12 Mar	Crater
15 Mar	Hydra
28 Mar	Corvus
28 Mar	Crux
30 Mar	Centaurus
30 Mar	Musca

April:

2 Apr	Coma Berenices
7 Apr	Canes Venatici
11 Apr	Virgo
30 Apr	Circinus

May:

2 May	Boötes
9 May	Libra
9 May	Lupus
13 May	Ursa Minor
19 May	Corona Borealis
19 May	Norma
21 May	Apus
23 May	Triangulum Australe
24 May	Draco

June:

3 Jun	Scorpius
6 Jun	Serpens
10 Jun	Ara
11 Jun	Ophiuchus
13 Jun	Hercules
30 Jun	Corona Australis

July:

1 Jul	Scutum
4 Jul	Lyra
7 Jul	Sagittarius
10 Jul	Telescopium
15 Jul	Pavo
16 Jul	Aquila
16 Jul	Sagitta
25 Jul	Vulpecula
30 Jul	Cygnus
31 Jul	Delphinus

August:

4 Aug	Microscopium
8 Aug	Capricornus

8 Aug	Equuleus	**November:**	
12 Aug	Indus	2 Nov	Fornax
25 Aug	Aquarius	7 Nov	Perseus
25 Aug	Piscis Austrinus	10 Nov	Eridanus
28 Aug	Grus	10 Nov	Horologium
28 Aug	Lacerta	19 Nov	Reticulum
		30 Nov	Taurus

September:

1 Sep	Pegasus
17 Sep	Tucana
26 Sep	Sculptor
27 Sep	Pisces
29 Sep	Cepheus

December:

1 Dec	Caelum
13 Dec	Orion
14 Dec	Lepus
14 Dec	Mensa
16 Dec	Pictor
17 Dec	Dorado
18 Dec	Columba
21 Dec	Auriga
23 Dec	Camelopardalis

October:

4 Oct	Phoenix
9 Oct	Andromeda
9 Oct	Cassiopeia
15 Oct	Cetus
23 Oct	Triangulum
26 Oct	Hydrus
30 Oct	Aries

Note: Octans is circumpolar and has no culmination date.

2 BY CONSTELLATION

Andromeda	9 Oct	Chamaeleon	1 Mar
Antlia	24 Feb	Circinus	30 Apr
Apus	21 May	Columba	18 Dec
Aquarius	25 Aug	Coma Berenices	2 Apr
Aquila	16 Jul	Corona Australis	30 Jun
Ara	10 Jun	Corona Borealis	19 May
Aries	30 Oct	Corvus	28 Mar
Auriga	21 Dec	Crater	12 Mar
Boötes	2 May	Crux	28 Mar
Caelum	1 Dec	Cygnus	30 Jul
Camelopardalis	23 Dec	Delphinus	31 Jul
Cancer	30 Jan	Dorado	17 Dec
Canes Venatici	7 Apr	Draco	24 May
Canis Major	2 Jan	Equuleus	8 Aug
Canis Minor	14 Jan	Eridanus	10 Nov
Capricornus	8 Aug	Fornax	2 Nov
Carina	31 Jan	Gemini	5 Jan
Cassiopeia	9 Oct	Grus	28 Aug
Centaurus	30 Mar	Hercules	13 Jun
Cepheus	29 Sep	Horologium	10 Nov
Cetus	15 Oct	Hydra	15 Mar

Hydrus	26 Oct	Pisces	27 Sep
Indus	12 Aug	Piscis Austrinus	25 Aug
Lacerta	28 Aug	Puppis	8 Jan
Leo	1 Mar	Pyxis	4 Feb
Leo Minor	23 Feb	Reticulum	19 Nov
Lepus	14 Dec	Sagitta	16 Jul
Libra	9 May	Sagittarius	7 Jul
Lupus	9 May	Scorpius	3 Jun
Lynx	19 Jan	Sculptor	26 Sep
Lyra	4 Jul	Scutum	1 Jul
Mensa	14 Dec	Serpens	6 Jun
Microscopium	4 Aug	Sextans	22 Feb
Monoceros	5 Jan	Taurus	30 Nov
Musca	30 Mar	Telescopium	10 Jul
Norma	19 May	Triangulum	23 Oct
Octans	—	Triangulum Australe	23 May
Ophiuchus	11 Jun	Tucana	17 Sep
Orion	13 Dec	Ursa Major	11 Mar
Pavo	15 Jul	Ursa Minor	13 May
Pegasus	1 Sep	Vela	13 Feb
Perseus	7 Nov	Virgo	11 Apr
Phoenix	4 Oct	Volans	18 Jan
Pictor	16 Dec	Vulpecula	25 Jul

The Moon and the planets in the constellations

In addition to the 12 'zodiacal' constellations, the Moon and planets may appear in the following constellations.

The Moon:

Auriga	Cetus	Corvus
Ophiuchus	Orion	Sextans

Total = 18

The Visible Planets:
(in addition to the constellations listed above)

Canis Minor	Crater	Hydra
Pegasus	Scutum	Serpens

Total = 24

Pluto:
(in addition to the constellations listed above)

Andromeda	Aquila	Boötes
Centaurus	Coma Berenices	Corona Australis
Equuleus	Eridanus	Leo Minor
Lupus	Lynx	Microscopium
Monoceros	Perseus	Piscis Austrinus
Sculptor	Triangulum	

Total = 41

Names of the constellations around the world

English	German	French	Italian
Andromeda	Andromeda f.	Andromède f.	Andromeda f.
Antlia	Luftpumpe f.	Machine Pneumatique f.	Macchina Pneumatica f.
Apus	Paradiesvogel m.	Oiseau de Paradis m.	Ucello del Paradiso m.
Aquarius	Wassermann m.	Verseau m.	Acquario m.
Aquila	Adler m.	Aigle m.	Aquila f.
Ara	Altar m.	Autel m.	Altare m.
Aries	Widder m.	Bélier m.	Ariete m.
Auriga	Fuhrmann m.	Cocher m.	Cocchiere m.
Boötes	Bärenhüter m.	Bouvier m.	Boote m.
Caelum	Grabstichel m.	Burin du Graveur m.	Bulino m.
Camelopardalis	Giraffe f.	Girafe f.	Giraffa f.
Cancer	Krebs m.	Écrivisse f.	Cancro m.
Canes Venatici	Jagdhunde mpl.	Chiens de Chasse mpl.	Levrieri mpl.
Canis Major	Grosser Hund m.	Grand Chien m.	Cane Maggiore m.
Canis Minor	Kleiner Hund m.	Petit Chien m.	Cane Minore m.
Capricornus	Steinbock m.	Capricorne m.	Capricorno m.
Carina	Kiel des Schiffes m.	Carène f.	Carena f.
Cassiopeia	Kassiopeia f.	Cassiopée f.	Cassiopea f.
Centaurus	Kentaur m.	Centaure m.	Centauro m.
Cepheus	Cepheus m.	Céphée m.	Cefeo m.
Cetus	Walfisch m.	Baleine f.	Balena f.
Chamaeleon	Chamaeleon n.	Caméléon m.	Camaleonte m.
Circinus	Zirkel m.	Compas m.	Compasso m.
Columba	Taube f.	Colombe f.	Colomba f.
Coma Berenices	Haar der Berenike n.	Chevelure de Bérénice f.	Chioma di Berenice f.
Corona Australis	Südliche Krone f.	Couronne Australe f.	Corona Australe f.
Corona Borealis	Nördliche Krone f.	Couronne Boréale f.	Corona Boreale f.
Corvus	Rabe m.	Corbeau m.	Corvo m.
Crater	Becher m.	Coupe f.	Cratere m.
Crux	Kreuz n.	Croiz du Sud f.	Croce del Sud f.
Cygnus	Schwan m.	Cygne m.	Cigno m.
Delphinus	Delphin m.	Dauphin m.	Delfino m.
Dorado	Schwertfisch m.	Dorade f.	Dorado m.
Draco	Drache m.	Dragon m.	Dragone m.
Equuleus	Pferdchen n.	Petit Cheval m.	Cavalluccio m.
Eridanus	Eridanusfluss m.	Eridan m.	Eridano m.
Fornax	Chemischer Ofen m.	Fourneau Chimique m.	Fornello Chimico m.
Gemini	Zwillinge mpl.	Gémeauz mpl.	Gemelli mpl.
Grus	Kranich m.	Grue f.	Gru f.

English	Spanish	Czech	Russian
Andromeda	Andrómeda f.	Andromeda f.	Ahndrawmyeda f.
Antlia	Bomba f.	Vyveva f.	Nassaws f.
Apus	Ave del Paraiso f.	Rajka f.	Raeeskaya pteetsa f.
Aquarius	Acuario m.	Vodnár m.	Vawdawlyeie m.
Aquila	Aguila f.	Orel m.	Awryell m.
Ara	Altar m.	Oltár m.	Zhertvyeneek m.
Aries	Carnero m.	Skopec m.	Awvyen m.
Auriga	Cochero m.	Vozka m.	Vawzneecheeie m.
Boötes	Boyero m.	Honák m.	Vawlawpahss m.
Caelum	Cielo m.	Rydio m.	Ryezyets m.
Camelopardalis	Jirafa f.	Zirafa f.	Zheeraf m.
Cancer	Cangrejo m.	Rak m.	Rak m.
Canes Venatici	Lebreles mpl.	Honici psi mpl.	Gawncheeyeh psih mpl.
Canis Major	Can Mayor m.	Velky pes m.	Bawlshawee pyes m.
Canis Minor	Can Menor m.	Maly pes m.	Maliy pyes m.
Capricornus	Capricornio m.	Kozorozec m.	Kawzyerawg m.
Carina	Carena f.	Kyl lodni m.	Keel m.
Cassiopeia	Casiopea f.	Kasiopeja f.	Kasseeawpeya f.
Centaurus	Centauro m.	Kentaur m.	Tsyentavr m.
Cepheus	Cefeo m.	Cefeus m.	Tsyefyeie m.
Cetus	Ballena f.	Velryba f.	Keet m.
Chamaeleon	Camaleón m.	Chameleon m.	Chamyelyeawn m.
Circinus	Brújula f.	Kruzitko n.	Tseerkool m.
Columba	Paloma f.	Holubice f.	Gawloob m.
Coma Berenices	Cabellara de Berenice f.	Vlas Berenicin m.	Vawlawsi vyerawneekee mpl.
Corona Australis	Corona Austral f.	Jizni koruna f.	Yuzhnaya Kawrawna f.
Corona Borealis	Corona Boreal f.	Severni koruna f.	Syevyernaya Kawrawna f.
Corvus	Cuervo m.	Havran m.	Vawrawn m.
Crater	Copa f.	Pohár m.	Chashcha f.
Crux	Cruz m.	Jizni Kriz m.	Kryest m.
Cygnus	Cisne m.	Labuf f.	Lyebyed m.
Delphinus	Delfin m.	Delfin m.	Dyelfeen m.
Dorado	Dorado m.	Mecoun m.	Zawlawtaya Riba f.
Draco	Dragón m.	Drak m.	Drakawn m.
Equuleus	Ecúleo m.	Konicek m.	Zhyeryebyenawk m.
Eridanus	Eridano m.	Eridanus m.	Ereedan m.
Fornax	Horno m.	Pec f.	Pyech f.
Gemini	Gemelos mpl.	Blizenci mpl.	Bleeznyetsi mpl.
Grus	Grulla f.	Jeráb m.	Zhooravl m.

67

English	German	French	Italian
Hercules	Herkules m.	Hercule m.	Ercole m.
Horologium	Pendeluhr f.	Horloge f.	Orologio m.
Hydra	Nördliche Wasserschlange f.	Hydre Femelle f.	Idra Femmina f.
Hydrus	Südliche Wasserschlange f.	Hydre Australe f.	Idra Australe f.
Indus	Indier m.	Indien m.	Indiano m.
Lacerta	Eidechse f.	Lézard m.	Lucertola f.
Leo	Löwe m.	Lion m.	Leone m.
Leo Minor	Kleiner Löwe m.	Petit Lion m.	Leoncino m.
Lepus	Hase m.	Lièvre m.	Lepre f.
Libra	Waage f.	Balance f.	Bilancia f.
Lupus	Wolf m.	Loup m.	Lupo m.
Lynx	Luchs m.	Lynx m.	Lince f.
Lyra	Leier f.	Lyre f.	Lira f.
Mensa	Tafelberg m.	Table f.	Mensa f.
Microscopium	Mikroskop n.	Microscope m.	Microscopio m.
Monoceros	Einhorn n.	Licorne f.	Unicorno m.
Musca	Fliege f.	Mouche f.	Mosca f.
Norma	Winkelmass n.	Equerre f.	Squadra f.
Octans	Oktant m.	Octant m.	Ottante m.
Ophiuchus	Schlangenträger m.	Serpentaire m.	Ofiuco m.
Orion	Himmelsjäger m.	Orion m.	Orione m.
Pavo	Pfau m.	Paon m.	Pavone m.
Pegasus	Pegasus m.	Pégase m.	Pegaso m.
Perseus	Perseus m.	Persée m.	Perseo m.
Phoenix	Phoenix m.	Phénix m.	Fenice f.
Pictor	Maler m.	Chevalet du Peintre m.	Pittore m.
Pisces	Fische mpl.	Poissons mpl.	Pesci mpl.
Piscis Austrinus	Südlicher Fisch m.	Poisson Austral m.	Pesce Australe m.
Puppis	Hinterteil des Schiffes m.	Poupe f.	Poppa f.
Pyxis	Schiffskompass m.	Boussole f.	Bussola f.
Reticulum	Netz n.	Réticule m.	Reticolo m.
Sagitta	Pfeil m.	Flèche f.	Saetta f.
Sagittarius	Schütze m.	Sagittaire m.	Sagittario m.
Scorpius	Skorpion m.	Scorpion m.	Scorpione f.
Sculptor	Bildhauer m.	Atelier du Sculpteur m.	Scultore m.
Scutum	Schild m.	Écu de Sobiesky m.	Scudo m.
Serpens	Schlange f.	Serpent m.	Serpente m.
Sextans	Sextant m.	Sextant m.	Sestante m.
Taurus	Stier m.	Taureau m.	Toro m.
Telescopium	Fernrohr n.	Télescope m.	Telescopio m.

English	Spanish	Czech	Russian
Hercules	Hercules m.	Herkules m.	Gyerkoolyess m.
Horologium	Reloj m.	Hodiny pl.	Chasi pl.
Hydra	Hidra f.	Vodni had m.	Geedra f.
Hydrus	Culebra f.	Maly vodni had m.	Yuzhnaya Geedra f.
Indus	Indio n.	Indián m.	Eendyeyets m.
Lacerta	Laguarto m.	Jesterka f.	Yashchyereetsa f.
Leo	León m.	Lev m.	Lyev m.
Leo Minor	León Menor m.	Maly lev m.	Maliy lyev m.
Lepus	Liebre f.	Zajic m.	Zayats m.
Libra	Balanza f.	Váhy pl.	Vyesi pl.
Lupus	Lobo m.	Vik m.	Vawlk m.
Lynx	Lince m.	Rys m.	Ris m.
Lyra	Lira f.	Lyra f.	Leera f.
Mensa	Mesa f.	Tabulová hora f.	Stawlawvaya Gawra f.
Microscopium	Microscopio m.	Drobnohled m.	Meekrowskawp m.
Monoceros	Unicornio f.	Jednorozec m.	Yedeenawrawg m.
Musca	Mosca f.	Moucha f.	Moohka f.
Norma	Escuadra f.	Pravitko n.	Naoogawlneek m.
Octans	Octante m.	Oktant m.	Awktant m.
Ophiuchus	Serpentario m.	Hadonos m.	Zmyeyenawsawts m.
Orion	Orión m.	Orión m.	Awreeawn m.
Pavo	Pavón m.	Páv m.	Pavleen m.
Pegasus	Pegaso m.	Pegas m.	Pyegas m.
Perseus	Perseo m.	Perseus m.	Pyersyeie m.
Phoenix	Fénix m.	Fénix m.	Fyeneeks m.
Pictor	Pintor m.	Malir m.	Zheevawpeesyets m.
Pisces	Peces mpl.	Ryby fpl.	Ribi fpl.
Piscis Austrinus	Pez Austral m.	Jizni ryba f.	Yuzhnaya riba f.
Puppis	Popa f.	Lodni zád f.	Kawrma f.
Pyxis	Compas m.	Kompas m.	Kawmpas m.
Reticulum	Reticulo m.	Sít f.	Syetka f.
Sagitta	Saeta f.	Síp m.	Stryela f.
Sagittarius	Sagitario m.	Strelec m.	Stryelyets m.
Scorpius	Escorpión m.	Stír m.	Skawrpeeawn m.
Sculptor	Escultor m.	Sochar m.	Skoolptawr m.
Scutum	Escudo m.	Stit m.	Shcheet m.
Serpens	Serpiente m.	Had m.	Zmyeya f.
Sextans	Sextante m.	Sextant m.	Syekstant m.
Taurus	Toro m.	Byk m.	Tyelyets m.
Telescopium	Telescopio m.	Dalekohled m.	Tyelyeskawp m.

English	German	French	Italian
Triangulum	Dreieck n.	Triangle m.	Triangolo m.
Triangulum Australe	Südliches Dreieck n.	Triangle Austral m.	Triangolo Australe m.
Tucana	Tukan m.	Toucan m.	Tucano m.
Ursa Major	Grosser Bär m.	Grande Ourse f.	Orsa Maggiore f.
Ursa Minor	Kleiner Bär m.	Petite Ourse f.	Orsa Minore f.
Vela	Segel des Schiffes n.	Voile f.	Vela f.
Virgo	Jungfrau f.	Vierge f.	Vergine f.
Volans	Fliegender Fisch m.	Poisson Volant m.	Pesce Volante m.
Vulpecula	Fuchs m.	Petit Renard m.	Volpetta f.

Note: m. masculine
f. feminine
n. neuter
pl. plural

English	Spanish	Czech	Russian
Triangulum	Triángulo m.	Trojúhelnik m.	Tryeoogawlneek m.
Triangulum Australe	Triángulo Austral m.	Jizni trojúhelnik m.	Yuzhnyie tryeoogawlneek m.
Tucana	Tucan m.	Tukan m.	Tookan m.
Ursa Major	Osa Mayor f.	Velky medved m.	Bawlshaya myedvyedeetsa f.
Ursa Minor	Osa Menor f.	Maly medved m.	Malaya myedvyedeetsa f.
Vela	Velas f.	Plachty fpl.	Paroosa mpl.
Virgo	Virgen n.	Panna f.	Dyeva f.
Volans	Pez Volandor m.	Létající ryba f.	Lyetoochaya riba f.
Vulpecula	Vulpeja m.	Liska f.	Leeseechka f.

The navigational stars

1 BY RIGHT ASCENSION

Designation	Proper name	Right ascension	Declination	Magnitude
α And	Alpheratz	00h07m	+29°00′	2.06
α Phe	Ankaa	00h26m	−42°26′	2.39
α Cas	Schedar	00h40m	+56°25′	2.23
β Cet	Diphda	00h43m	−18°05′	2.04
α Eri	Achernar	01h37m	−57°19′	0.46
α Ari	Hamal	02h06m	+23°23′	2.00
θ Eri	Acamar	02h57m	−40°22′	2.91
α Cet	Menkar	03h02m	+4°02′	2.53
α Per	Mirfak	03h23m	+49°48′	1.79
α Tau	Aldebaran	04h35m	+16°28′	0.85
β Ori	Rigel	05h13m	−8°13′	0.12
α Aur	Capella	05h15m	+45°59′	0.08
γ Ori	Bellatrix	05h24m	+6°20′	1.64
β Tau	El Nath	05h25m	+28°35′	1.65
ε Ori	Alnilam	05h35m	−1°12′	1.70
α Ori	Betelgeuse	05h54m	+7°24′	0.50
α Car	Canopus	06h23m	−52°41′	−0.72
α CMa	Sirius	06h44m	−16°41′	−1.46
ε CMa	Adhara	06h58m	−28°57′	1.50
α CMi	Procyon	07h38m	+5°16′	0.38
β Gem	Pollux	07h44m	+28°04′	1.14
ε Car	Avior	08h22m	−59°27′	1.86
λ Vel	Alsuhail	09h07m	−43°22′	2.21
β Car	Miaplacidus	09h13m	−69°38′	1.68
α Hya	Alphard	09h27m	−8°35′	1.98
α Leo	Regulus	10h07m	+12°03′	1.35
α UMa	Dubhe	11h02m	+61°50′	1.79
β Leo	Denebola	11h48m	+14°40′	2.14
α Cru	Acrux	12h25m	−63°00′	0.90
γ Cru	Gacrux	12h30m	−57°01′	1.63
ε UMa	Alioth	12h53m	+56°03′	1.77
α Vir	Spica	13h24m	−11°04′	0.97
η UMa	Alkaid	13h47m	+49°26′	1.86
β Cen	Hadar	14h03m	−60°16′	0.61
θ Cen	Menkent	14h05m	−36°17′	2.06
α Boo	Arcturus	14h14m	+19°16′	−0.04
α Cen	Rigil Kentaurus	14h38m	−60°46′	−0.27
α Lib	Zubenelgenubi	14h50m	−15°57′	2.75

Designation	Proper name	Right ascension	Declination	Magnitude
β UMi	Kochab	14h50m	+74°13′	2.08
α CrB	Alphecca	15h34m	+26°46′	2.23
α Sco	Antares	16h28m	−26°23′	0.96
α TrA	Atria	16h47m	−69°00′	1.92
η Oph	Sabik	17h09m	−15°42′	2.43
λ Sco	Shaula	17h32m	−37°05′	1.63
α Oph	Rasalhague	17h34m	+12°34′	2.08
γ Dra	Eltanin	17h56m	+51°29′	2.23
ε Sgr	Kaus Australis	18h23m	−34°23′	1.85
α Lyr	Vega	18h36m	+38°46′	0.03
σ Sgr	Nunki	18h54m	−26°19′	2.02
α Aql	Altair	19h50m	+8°49′	0.77
α Pav	Peacock	20h24m	−56°47′	1.94
α Cyg	Deneb	20h41m	+45°13′	1.25
ε Cyg	Gienah	20h46m	+33°57′	2.46
ε Peg	Enif	21h43m	+9°48′	2.39
α Gru	Al Na'ir	22h07m	−47°02′	1.74
α PsA	Fomalhaut	22h56m	−29°43′	1.16
α Peg	Markab	23h04m	+15°06′	2.49
α UMi	Polaris	02h14m	+89°11′	2.02

2 IN ALPHABETICAL ORDER

Designation	Proper name	Right ascension	Declination	Magnitude
θ Eri	Acamar	02h57m	−40°22′	2.91
α Eri	Achernar	01h37m	−57°19′	0.46
α Cru	Acrux	12h25m	−63°00′	0.90
ε CMa	Adhara	06h58m	−28°57′	1.50
α Tau	Aldebaran	04h35m	+16°28′	0.85
ε UMa	Alioth	12h53m	+56°03′	1.77
η UMa	Alkaid	13h47m	+49°26′	1.86
α Gru	Alnair	22h07m	−47°02′	1.74
ε Ori	Alnilam	05h35m	−1°12′	1.70
α Hya	Alphard	09h27m	−8°35′	1.98
α CrB	Alphecca	15h34m	+26°46′	2.23
α And	Alpheratz	00h07m	+29°00′	2.06
λ Vel	Alsuhail	09h07m	−43°22′	2.21
α Aql	Altair	19h50m	+8°49′	0.77
α Phe	Ankaa	00h26m	−42°26′	2.39
α Sco	Antares	16h28m	−26°23′	0.96

Designation	Proper name	Right ascension	Declination	Magnitude
α Boo	Arcturus	14h14m	+19°16′	−0.04
α TrA	Atria	16h47m	−69°00′	1.92
ε Car	Avior	08h22m	−59°27′	1.86
γ Ori	Bellatrix	05h24m	+6°20′	1.64
α Ori	Betelgeuse	05h54m	+7°24′	0.50
α Car	Canopus	06h23m	−52°41′	−0.72
α Aur	Capella	05h15m	+45°59′	0.08
α Cyg	Deneb	20h41m	+45°13′	1.25
β Leo	Denebola	11h48m	+14°40′	2.14
β Cet	Diphda	00h43m	−18°05′	2.04
α UMa	Dubhe	11h02m	+61°50′	1.79
β Tau	El Nath	05h25m	+28°35′	1.65
γ Dra	Eltanin	17h56m	+51°29′	2.23
ε Peg	Enif	21h43m	+9°48′	2.39
α PsA	Fomalhaut	22h56m	−29°43′	1.16
γ Cru	Gacrux	12h30m	−57°01′	1.63
ε Cyg	Gienah	20h46m	+33°57′	2.46
β Cen	Hadar	14h03m	−60°16′	0.61
α Ari	Hamal	02h06m	+23°23′	2.00
ε Sgr	Kaus Australis	18h23m	−34°23′	1.85
β UMi	Kochab	14h50m	+74°13′	2.08
α Peg	Markab	23h04m	+15°06′	2.49
α Cet	Menkar	03h02m	+4°02′	2.53
θ Cen	Menkent	14h05m	−36°17′	2.06
β Car	Miaplacidus	09h13m	−69°38′	1.68
α Per	Mirfak	03h23m	+49°48′	1.79
σ Sgr	Nunki	18h54m	−26°19′	2.02
α Pav	Peacock	20h24m	−56°47′	1.94
β Gem	Pollux	07h44m	+28°04′	1.14
α CMi	Procyon	07h38m	+5°16′	0.38
α Oph	Rasalhague	17h34m	+12°34′	2.08
α Leo	Regulus	10h07m	+12°03′	1.35
β Ori	Rigel	05h13m	−8°13′	0.12
α Cen	Rigil Kentaurus	14h38m	−60°46′	−0.27
η Oph	Sabik	17h09m	−15°42′	2.43
α Cas	Schedar	00h40m	+56°25′	2.23
λ Sco	Shaula	17h32m	−37°05′	1.63
α CMa	Sirius	06h44m	−16°41′	−1.46
α Vir	Spica	13h24m	−11°04′	0.97
α Lyr	Vega	18h36m	+38°46′	0.03
α Lib	Zubenelgenubi	14h50m	−15°57′	2.75
α UMi	Polaris	02h14m	+89°11′	2.02

The 200 nearest stars

Rank	Designation	Constellation	Apparent magnitude	Absolute magnitude	Parallax (arcsec)	Distance (light years)
	The Sun	—	−26.7	4.9	—	0.000016
1	Proxima Centauri	Cen	10.7	15.1	.761	4.28
2	α Cen A	Cen	−0.3	4.6	.743	4.39
	α Cen B	Cen	1.4	5.8	.743	4.39
3	Barnard's Star	Oph	9.5	13.2	.548	5.95
4	Wolf 359	Leo	13.5	16.5	.429	7.60
5	Lalande 21185	UMa	7.5	10.5	.396	8.23
6	Sirius A	CMa	−1.5	0.7	.377	8.65
	Sirius B	CMa	8.5	11.4	.377	8.65
7	UV Cet A	Cet	12.5	15.3	.367	8.88
	UV Cet B	Cet	13.0	15.8	.367	8.88
8	Ross 154	Sgr	10.6	13.3	.345	9.45
9	Ross 248	And	12.2	14.7	.318	10.25
10	ε Eri	Eri	3.7	6.1	.305	10.69
11	L 789–6	Aqr	12.2	14.6	.302	10.79
12	Ross 128	Vir	11.1	13.5	.301	10.83
13	61 Cyg A	Cyg	5.2	7.5	.296	11.01
	61 Cyg B	Cyg	6.0	8.4	.296	11.01
14	ε Ind	Ind	4.7	7.0	.291	11.20
15	Procyon A	CMi	0.4	2.8	.287	11.36
	Procyon B	CMi	10.8	13.1	.287	11.36
16	LFT 1431	Dra	8.9	9.7	.284	11.48
	LFT 1432	Dra	9.7	11.9	.284	11.48
17	Groombridge 34 A	And	8.1	10.3	.282	11.56
	Groombridge 34 B	And	11.0	13.3	.282	11.56
18	τ Cet	Cet	3.5	5.7	.279	11.68
19	Lacaille 9352	PsA	7.4	9.6	.277	11.77
20	BD+5°1668	CMi	9.8	11.7	.270	12.07
21	Cordoba 29191	Mic	6.7	9.0	.260	12.54
22	Kapteyn's Star	Pic	8.8	10.8	.256	12.73
23	Krueger 60 A	Cep	9.8	11.8	.253	12.89
	Krueger 60 B	Cep	11.4	13.4	.253	12.89
24	Ross 614 A	Mon	11.1	13.1	.252	12.94
	Ross 614 B	Mon	14.4	16.4	.252	12.94
25	Wolf 1061	Oph	10.1	12.1	.249	13.09
26	van Maanen's Star	Psc	12.4	14.3	.239	13.64
27	Wolf 424 A	Vir	12.5	14.3	.231	14.11
	Wolf 424 B	Vir	13.4	15.2	.231	14.11
28	BD+44°4548	Scl	8.6	10.3	.225	14.49

Rank	Designation	Constellation	Apparent magnitude	Absolute magnitude	Parallax (arcsec)	Distance (light years)
29	BD+50°1725	UMa	6.6	8.3	.222	14.68
30	LFT 1351	Ara	9.4	11.0	.216	15.09
31	BD+68°946	Dra	9.2	10.8	.214	15.23
32	LFT 1640	Gru	8.7	10.3	.214	15.23
33	LFT 1358	Sco	11.2	12.8	.213	15.31
34	LFT 171	Ari	12.3	13.9	.213	15.31
35	Ross 780	Aqr	10.1	11.7	.209	15.60
36	L 145–141	Mus	11.5	13.1	.206	15.83
37	AD Leo	Leo	9.4v[a]	11.0v	.206	15.83
38	40 Eri A	Eri	4.4	6.0	.205	15.90
	40 Eri B	Eri	9.5	11.1	.205	15.90
	40 Eri C	Eri	11.2	12.7	.205	15.90
39	Wolf 498	Boo	8.5	10.0	.205	15.90
40	Altair	Aql	0.8	2.1	.196	16.63
41	LFT 849	Cam	10.9	12.4	.195	16.72
42	70 Oph A	Oph	4.0	5.5	.195	16.72
	70 Oph B	Oph	6.0	7.5	.195	16.72
43	EV Lac	Lac	10.0v	11.5v	.195	16.72
44	WX UMa A	UMa	8.8	10.2	.192	16.98
	WX UMa B	UMa	14.5v	16.0v	.192	16.98
45	36 Oph A	Oph	4.4	5.7	.189	17.25
	36 Oph B	Oph	5.1	6.5	.189	17.25
46	ADS 9446 A	Lib	5.8	7.1	.180	18.11
	ADS 9446 B	Lib	8.0	9.3	.180	18.11
47	LFT 1332	Oph	6.3	7.6	.178	18.31
48	LFT 1529	Sgr	5.3	6.5	.177	18.42
	LFT 1530	Sgr	11.5	12.7	.177	18.42
49	σ Dra	Dra	4.7	5.9	.176	18.52
50	LFT 1469	Sgr	13.7	14.9	.175	18.63
51	δ Pav	Pav	3.6	4.8	.175	18.63
52	BD+1°4774	Psc	9.0	10.2	.174	18.74
53	BD–21°1377	Lep	8.1	9.3	.174	18.74
54	L 97–12	Vol	14.5	15.7	.173	18.84
55	η Cas A	Cas	3.4	4.6	.172	18.95
	η Cas B	Cas	7.5	8.7	.172	18.95
56	Wolf 1055 A	Aql	9.1	10.3	.172	18.95
	van Biesbroeck's Star	Aql	17.5v	18.7v	.172	18.95
57	LFT 571	Pup	13.8	15.0	.171	19.06
58	LFT 1372	Pav	12.9	14.1	.170	19.18
59	Wolf 1453	Ori	8.0	9.1	.170	19.18
60	Ross 986	Aur	11.5	12.6	.169	19.29
61	LFT 1208	Lup	10.1	11.2	.169	19.29
62	Wolf 629	Oph	11.7	12.8	.169	19.29

Rank	Designation	Constellation	Apparent magnitude	Absolute magnitude	Parallax (arcsec)	Distance (light years)
63	Wolf 294	Gem	9.9	11.0	.168	19.40
64	YZ CMi	CMi	11.2v	12.3v	.167	19.52
65	Ross 47	Ori	11.6	12.7	.166	19.64
66	LP 658–2	Ori	14.5	15.6	.166	19.64
67	LFT 634	UMa	7.6	8.7	.166	19.64
	LFT 635	UMa	7.7	8.8	.166	19.64
68	LFT 1532	Sgr	8.0	9.0	.164	19.88
69	82 Eri	Eri	4.3	5.3	.161	20.25
70	BD–11°3759	Lib	11.4	12.4	.160	20.38
71	β Hyi	Hyi	2.8	3.8	.159	20.50
72	V1054 Oph A	Oph	9.0	10.9	.156	20.90
	V1054 Oph B	Oph	9.8	10.8	.156	20.90
	V1054 Oph C	Oph	16.7	17.6	.156	20.90
73	EQ Peg A	Peg	10.4	11.3	.155	21.03
	EQ Peg B	Peg	12.5v	13.5v	.155	21.03
74	Ross 775	Peg	10.4	11.3	.154	21.17
75	LFT 1326	Her	9.4	10.3	.153	21.31
	LFT 1327	Her	10.3	11.2	.153	21.31
76	Wolf 562	Lib	10.6	11.5	.153	21.31
77	p Eri A	Eri	5.1	6.0	.153	21.31
	p Eri B	Eri	5.9	6.8	.153	21.31
78	Ross 619	Cnc	12.8	13.7	.151	21.59
79	Ross 104	Leo	10.0	10.9	.151	21.59
80	Fomalhaut	PsA	1.2	1.8	.149	21.88
81	ξ Boo A	Boo	4.5	5.4	.148	22.03
	ξ Boo B	Boo	6.9	7.8	.148	22.03
82	BD+56°2966	Cas	5.6	6.4	.147	22.18
83	Ross 671	Peg	8.7	9.5	.146	22.33
84	BD+4°123	Psc	5.8	6.5	.144	22.64
85	BD+6°398 A	Cet	5.8	6.6	.144	22.64
	BD+6°398 B	Cet	11.7	12.4	.144	22.64
86	L 745–46 A	Pup	13.0	13.8	.142	22.96
	L 745–46 B	Pup	17.6	18.4	.142	22.96
87	Wolf 358	Leo	11.7	12.4	.142	22.96
88	HD 156384 A	Sco	5.9	6.6	.140	23.29
	HD 156384 B	Sco	7.2	7.9	.140	23.29
	HD 156384 C	Sco	10.2	10.9	.140	23.29
89	ζ Tuc	Tuc	4.2	5.0	.139	23.45
90	LP 425–140	Cnc	18.6	19.3	.139	23.45
91	BD+61°2068	Cep	8.5	9.2	.139	23.45
92	LFT 1273	Her	10.3	11.0	.138	23.62
93	BD–3°4233	Ser	9.4	10.1	.138	23.62
94	Wolf 489	Vir	14.7	15.4	.135	24.15

Rank	Designation	Constellation	Apparent magnitude	Absolute magnitude	Parallax (arcsec)	Distance (light years)
95	107 Psc	Psc	5.2	5.9	.134	24.33
96	Wolf 922	Cap	12.0	12.6	.134	24.33
97	LFT 215	Ari	14.6	15.2	.133	24.51
98	π^3 Ori	Ori	3.2	3.8	.132	24.70
99	Wolf 718	Oph	7.5	8.1	.132	24.70
100	BD–21°1051 A	Lep	8.3	8.9	.131	24.89
	BD–21°1051 B	Lep	10.7	11.3	.131	24.89
101	Ross 490	Vir	9.0	9.6	.131	24.89
102	LFT 1266 (A)	Sco	12.0	12.6	.131	24.89
	LFT 1267 (B)	Sco	16.0	16.6	.131	24.89
103	41 Ara A	Ara	5.5	6.1	.131	24.89
	41 Ara B	Ara	8.7	9.3	.131	24.89
104	Ross 446	Sex	9.6	10.2	.130	25.08
105	LFT 1754	Aqr	7.9	8.5	.130	25.08
106	χ Dra	Dra	3.6	4.1	.129	25.27
107	μ Her A	Her	3.4	4.0	.128	25.47
	μ Her B	Her	9.8	10.3	.128	25.47
	μ Her C	Her	10.7	11.2	.128	25.47
108	LTT 9283	PsA	6.5	7.0	.128	25.47
109	LFT 1552	Dra	10.4	11.0	.127	25.67
110	ξ UMa A	UMa	3.8	4.3	.127	25.67
	ξ UMa B	UMa	4.8	5.3	.127	25.67
111	BD–13°544	Eri	6.1	6.6	.127	25.67
112	μ Cas	Cas	5.1	5.6	.127	25.67
113	LFT 1363	Her	9.6	10.1	.126	25.87
114	Ross 556	Ari	10.6	11.1	.125	26.08
115	Vega	Lyr	0.0	0.3	.124	26.29
116	γ Lep A	Lep	3.6	4.1	.123	26.50
	γ Lep B	Lep	6.2	6.6	.123	26.50
	γ Lep C	Lep	16.1	16.5	.123	26.50
117	LFT 661	Hya	12.7	13.1	.122	26.72
118	17 Lyr C A	Lyr	11.8	12.2	.122	26.72
	17 Lyr C B	Lyr	12.1	12.5	.122	26.72
119	LTT 8181	Mic	10.1	10.6	.122	26.72
	LTT 8182	Mic	11.1	11.5	.122	26.72
120	LFT 1747	Oct	11.7	12.1	.122	26.72
121	L 362–81	Phe	13.1	13.5	.122	26.72
122	LFT 930	Cen	10.7	11.1	.121	26.94
123	LFT 839	Cen	10.9	11.3	.120	27.17
124	β Com	Com	4.3	4.7	.120	27.17
125	LFT 1535	Cap	5.7	6.1	.120	27.17
126	SZ UMa	UMa	9.3	9.7	.119	27.39
127	61 Vir	Vir	4.7	5.1	.119	27.39

Rank	Designation	Constellation	Apparent magnitude	Absolute magnitude	Parallax (arcsec)	Distance (light years)
128	Ross 730	Sge	10.8	11.2	.119	27.39
	Ross 731	Sge	10.8	11.1	.119	27.39
129	BD+53°935	Aur	9.8	10.1	.117	27.86
130	Ross 64	Gem	13.3	13.6	.117	27.86
131	LFT 598	Pyx	13.2	13.5	.116	28.10
132	LFT 698	Sex	10.8	11.1	.116	28.10
133	LTT 13665	Dra	10.8	11.1	.116	28.10
134	γ Pav	Pav	4.2	4.5	.116	28.10
135	Ross 41	Ori	12.5	12.8	.115	28.35
136	α Men	Men	5.1	5.4	.115	28.35
137	LFT 502 A	Pup	10.8	11.1	.115	28.35
	LFT 502 B	Pup	11.7	12.0	.115	28.35
138	Ross 318	Cas	10.1	10.3	.114	28.60
139	LFT 1851	Psc	11.0	11.2	.114	28.60
140	BD+63°238	Cas	5.6	5.9	.114	28.60
141	BD−18°359	Cet	10.2	10.5	.114	28.60
142	LFT 117	Tuc	11.2	11.5	.113	28.85
143	L 532−81	Pyx	12.0	12.3	.113	28.85
144	Ross 905	Leo	10.7	11.0	.113	28.85
145	Wolf 1329	Aqr	10.4	10.7	.112	29.11
146	V388 Cas	Cas	13.7	13.9	.111	29.37
147	LFT 193	For	10.2	10.4	.111	29.37
148	δ Eri	Eri	3.5	3.7	.111	29.37
149	Wolf 437	Vir	11.4	11.6	.111	29.37
150	LFT 1218	Lib	11.7	11.9	.111	29.37
151	BD−12°2918 A	Hya	10.1	10.3	.110	29.64
	BD−12°2918 B	Hya	10.8	11.0	.110	29.64
152	61 UMa	UMa	5.3	5.6	.110	29.64
153	Groombridge 1830	UMa	6.5	6.7	.110	29.64
154	Ross 695	Crv	11.7	11.9	.110	29.64
155	Wolf 46	Cas	9.6	9.8	.109	29.91
156	BD−5°1123	Eri	6.2	6.4	.109	29.91
157	11 LMi A	LMi	5.4	5.6	.109	29.91
	11 LMi B	LMi	13.0	13.2	.109	29.91
158	LFT 682	Vel	10.6	10.8	.109	29.91
159	LFT 709	Sex	11.2	11.4	.109	29.91
160	β CVn	CVn	4.3	4.5	.109	29.91
161	BD−21°6267 A	Aqr	9.1	9.3	.109	29.91
	BD−21°6267 B	Aqr	11.4	11.6	.109	29.91
162	Wolf 457	Vir	15.9	16.1	.108	30.19
163	χ Cet	Cet	4.8	5.0	.107	30.47
164	LTT 8214	Mic	8.6	8.8	.107	30.47
165	BD−3°2870	Sex	9.3	9.4	.105	31.05

Rank	Designation	Constellation	Apparent magnitude	Absolute magnitude	Parallax (arcsec)	Distance (light years)
166	LFT 823 A	Hya	6.0	6.1	.105	31.05
	LFT 823 B	Hya	15.0	15.1	.105	31.05
167	LFT 1297	Ara	14.4	14.5	.105	31.05
168	BD+0°4810	Aqr	9.2	9.3	.105	31.05
169	BD+10°1032 A	Ori	10.4	10.5	.104	31.35
	BD+10°1032 B	Ori	12.4	12.5	.104	31.35
170	Wolf 287	Gem	9.6	9.7	.104	31.35
171	BD–5°1844 A	Mon	6.7	6.8	.104	31.35
	BD–5°1844 B	Mon	10.1	10.2	.104	31.35
172	LFT 643	Car	11.3	11.4	.104	31.35
173	LFT 1088	Cen	6.7	6.8	.104	31.35
174	ζ Her A	Her	2.8	2.9	.104	31.35
	ζ Her B	Her	5.5	5.6	.104	31.35
175	BD+33°2777	Her	8.1	8.2	.104	31.35
176	Wolf 751	Oph	9.3	9.4	.104	31.35
177	LTT 7658	Aql	12.1	12.2	.104	31.35
	LTT 7659	Aql	12.3	12.4	.104	31.35
178	LFT 1699	Aqr	13.5	13.6	.104	31.35
	LFT 1700	Aqr	14.4	14.5	.104	31.35
179	LFT 445	Cam	10.5	10.6	.103	31.65
180	LFT 729	Sex	12.7	12.8	.103	31.65
181	BD+36°2393	CVn	9.0	9.1	.103	31.65
182	LFT 1371	Her	10.5	10.6	.103	31.65
183	LFT 1813	Oct	7.1	7.2	.103	31.65
184	LTT 1830	Eri	8.1	8.1	.102	31.96
	LTT 1831	Eri	11.5	11.5	.102	31.96
185	LFT 865	Hya	7.0	7.0	.102	31.96
186	η Boo	Boo	2.7	2.9	.102	31.96
187	Ross 863	Her	11.6	11.6	.102	31.96
188	χ¹ Ori	Ori	4.4	4.4	.101	32.27
189	LFT 764	Crt	11.8	11.8	.101	32.27
190	Wolf 636	Oph	10.1	10.1	.101	32.27
191	Ross 165 A	Vul	12.7	12.7	.101	32.27
	Ross 165 B	Vul	13.7	13.7	.101	32.27
192	54 Psc	Psc	5.9	5.9	.100	32.60
193	Wolf 124	Cet	10.0	10.0	.100	32.60
194	LTT 4204	Crt	8.7	8.7	.100	32.60
	LTT 4205	Crt	11.0	11.0	.100	32.60
195	β Vir	Vir	3.6	3.6	.100	32.60
196	BD+46°1889	CVn	10.0	10.0	.100	32.60
197	Ross 594	Per	13.7	13.7	.099	32.93
198	γ Vir A	Vir	2.7	2.7	.099	32.93
	γ Vir B	Vir	3.5	3.5	.099	32.93

Rank	Designation	Constellation	Apparent magnitude	Absolute magnitude	Parallax (arcsec)	Distance (light years)
199	BD+35°2436 A	CVn	9.5	9.5	.099	32.93
	BD+35°2436 B	CVn	12.1	12.1	.099	32.93
200	LFT 1338	Aps	7.5	7.5	.099	32.93

[a] v indicates variable

The 'new' constellations
(created since the time of Ptolemy)

by Johannes Hevelius (1687):

Canes Venatici	Lacerta	Leo Minor
Lynx	Scutum	Sextans
Vulpecula		

by Pieter Dirksz Keyser and Frederick de Houtman (1596):

Apus	Chamaeleon	Dorado
Grus	Hydrus	Indus
Musca	Pavo	Phoenix
Tucana	Volans	

by Nicolas Louis de Lacaille (1756):

Antlia	Caelum	Carina
Circinus	Fornax	Horologium
Mensa	Microscopium	Norma
Octans	Pictor	Puppis
Pyxis	Reticulum	Sculptor
Telescopium	Vela	

by Gerard Mercator (1551):

Coma Berenices

by Petrus Plancius (1592, 1613):

Camelopardalis (1613)	Columba (1592)	Monoceros (1613)

by Amerigo Vespucci (1503):

Crux	Triangulum Australe

The 'original' 48 constellations

1 THE LIST OF EUDOXUS OF KNIDOS

As no work of Eudoxus survives these are taken from the work of Aratos.

Northern constellations:

The Lesser Bear (Ursa Minor)
The Greater Bear (Ursa Major)
The Bearward or Ploughman (Boötes)
The Serpent (Draco)
Kêpheus (Cepheus)
Kassiepeia (Cassiopeia)
Andromeda (Andromeda)
Perseus (Perseus)
The Delta-Shaped Figure (Triangulum)
The Horse (Pegasus)

The Dolphin (Delphinus)
The Charioteer (Auriga)
The Kneeler (Hercules)
The Lyre (Lyra)
The Bird (Cygnus)
The Eagle (Aquila)
The Arrow (Sagitta)
The Crown (Corona Borealis)
The Snake-Holder (Ophiuchus)

Central or zodiacal constellations:

The Ram (Aries)
The Bull (Taurus)
The Twins (Gemini)
The Crab (Cancer)
The Lion (Leo)
The Virgin (Virgo)
The Claws (Libra)

The Scorpion (Scorpius)
The Archer (Sagittarius)
The Goat (Capricornus)
The Water-Pourer (Aquarius)
The Fishes (Pisces)
The Clusterers (The Pleiades)

Southern constellations:

Orîôn (Orion)
The Dog (Canis Major)
The Hare (Lepus)
Argô (The Ship)
The Sea-Monster (Cetus)
The Stream (Eridanus)

The Fish (Piscis Austrinus)
The Altar (Ara)
The Centaur (Centaurus)
The Water-Snake (Hydra)
The Bowl (Crater)
The Crow (Corvus)

Note: Aratos notices, but does not name, The Southern Crown (Corona Australis). He also mentions five individual stars: Bear-watcher (Arktouros – Arcturus), Ear-of-corn (Stachys – Spica), Fruit-plucking-herald (Protrygêtêr – Vindemiatrix), Scorcher (Seirios – Sirius), and Dog's-precursor (Prokyôn – Procyon).

2 THE LIST OF PTOLEMY

Northern figures:

The Little Bear (Ursa Minor)
The Great Bear (Ursa Major)
The Serpent (Draco)

Kephêus (Cepheus)
The Ploughman (Boötes)
The Northern Crown (Corona Borealis)
The Kneeler (Hercules)
The Lyre (Lyra)
The Bird (Cygnus)
Kassiepeia (Cassiopeia)
Perseus (Perseus)

The Charioteer (Auriga)
The Snake-Holder (Ophiuchus)
The Snake of the Snake-Holder
  (Serpens)
The Arrow (Sagitta)
The Eagle (Aquila)
The Dolphin (Delphinus)
The Foremost-Part of a Horse
  (Equuleus)
The Horse (Pegasus)
Andromeda (Andromeda)
The Triangle (Triangulum)

The northern figures in the zodiac:

The Ram (Aries)
The Bull (Taurus)
The Twins (Gemini)

The Crab (Cancer)
The Lion (Leo)
The Virgin (Virgo)

The Southern figures in the zodiac:

The Claws (Libra)
The Scorpion (Scorpius)
The Archer (Sagittarius)

Capricorn (Capricornus)
The Water-pourer (Aquarius)
The Fishes (Pisces)

Southern figures:

The Sea-Monster (Cetus)
Orîôn (Orion)
The Stream (Eridanus)
The Hare (Lepus)
The Dog (Canis Major)
The Fore-Dog (Canis Minor)
Argô (Argo Navis)
The Water-Snake (Hydra)

The Bowl (Crater)
The Crow (Corvus)
The Centaur (Centaurus)
The Wild-Beast (Lupus)
The Censer (Ara)
The Southern Crown (Corona
  Australis)
The Southern Fish (Piscis
  Austrinus)

Overall brightness of the constellations

1 IN ORDER OF BRIGHTNESS

Rank	Constellation	Overall brightness
1	Crux	29.218
2	Corona Australis	16.446
3	Carina	15.581
4	Vela	15.211
5	Lupus	14.984
6	Canis Major	14.733
7	Puppis	13.810
8	Musca	13.732
9	Orion	12.960
10	Scorpius	12.480
11	Corona Borealis	12.310
12	Taurus	12.292
13	Lacerta	11.460
14	Triangulum Australe	10.911
15	Vulpecula	10.814
16	Circinus	10.712
17	Perseus	10.569
18	Sagitta	10.009
19	Volans	9.904
20	Chamaeleon	9.879
21	Cygnus	9.826
22	Cepheus	9.697
23	Reticulum	9.654
24	Lepus	9.646
25	Centaurus	9.525
26	Gemini	9.148
27	Triangulum	9.101
28	Lyra	9.076
29	Columba	8.883
30	Cassiopeia	8.523
31	Norma	8.470
32	Dorado	8.372
33	Scutum	8.249
34	Ara	8.015
35	Sagittarius	7.493
36	Capricornus	7.489
37	Andromeda	7.476
38	Monoceros	7.476

Rank	Constellation	Overall brightness
39	Pavo	7.414
40	Draco	7.295
41	Aquila	7.203
42	Microscopium	7.160
43	Auriga	7.149
44	Canis Minor	7.089
45	Ursa Minor	7.035
46	Equuleus	6.979
47	Eridanus	6.942
48	Hercules	6.938
49	Telescopium	6.759
50	Grus	6.566
51	Libra	6.505
52	Leo Minor	6.467
53	Aries	6.344
54	Piscis Austrinus	6.113
55	Pictor	6.080
56	Corvus	5.985
57	Coma Berenices	5.951
58	Camelopardalis	5.946
59	Boötes	5.845
60	Octans	5.841
61	Delphinus	5.834
62	Ophiuchus	5.800
63	Hydrus	5.760
64	Phoenix	5.753
65	Aquarius	5.715
66	Lynx	5.684
67	Serpens	5.652
68	Pisces	5.622
69	Ursa Major	5.548
70	Leo	5.491
71	Hydra	5.449
72	Pyxis	5.434
73	Mensa	5.212
74	Tucana	5.092
75	Pegasus	5.086
76	Apus	4.847
77	Cetus	4.710
78	Cancer	4.545
79	Virgo	4.481
80	Indus	4.422
81	Horologium	4.018
82	Crater	3.895

Rank	Constellation	Overall brightness
83	Antlia	3.767
84	Canes Venatici	3.224
85	Caelum	3.204
86	Sculptor	3.159
87	Fornax	3.019
88	Sextans	1.595

2 BY CONSTELLATION

Constellation	Rank	Overall brightness
Andromeda	37	7.476
Antlia	83	3.767
Apus	76	4.847
Aquarius	65	5.715
Aquila	41	7.203
Ara	34	8.015
Aries	53	6.344
Auriga	43	7.149
Boötes	59	5.845
Caelum	85	3.204
Camelopardalis	58	5.946
Cancer	78	4.545
Canes Venatici	84	3.224
Canis Major	6	14.733
Canis Minor	44	7.089
Capricornus	36	7.489
Carina	3	15.581
Cassiopeia	30	8.523
Centaurus	25	9.525
Cepheus	22	9.697
Cetus	77	4.710
Chamaeleon	20	9.879
Circinus	16	10.712
Columba	29	8.883
Coma Berenices	57	5.951
Corona Australis	2	16.446
Corona Borealis	11	12.310
Corvus	56	5.985
Crater	82	3.895
Crux	1	29.218

Constellation	Rank	Overall brightness
Cygnus	21	9.826
Delphinus	61	5.834
Dorado	32	8.372
Draco	40	7.295
Equuleus	46	6.979
Eridanus	47	6.942
Fornax	87	3.019
Gemini	26	9.148
Grus	50	6.566
Hercules	48	6.938
Horologium	81	4.018
Hydra	71	5.449
Hydrus	63	5.760
Indus	80	4.422
Lacerta	13	11.460
Leo	70	5.491
Leo Minor	52	6.467
Lepus	23	9.646
Libra	51	6.505
Lupus	5	14.984
Lynx	66	5.684
Lyra	28	9.076
Mensa	73	5.212
Microscopium	42	7.160
Monoceros	38	7.476
Musca	8	13.732
Norma	31	8.470
Octans	60	5.841
Ophiuchus	62	5.800
Orion	9	12.960
Pavo	39	7.414
Pegasus	75	5.086
Perseus	17	10.569
Phoenix	64	5.753
Pictor	55	6.080
Pisces	68	5.622
Piscis Austrinus	54	6.113
Puppis	7	13.810
Pyxis	72	5.434
Reticulum	24	9.654
Sagitta	18	10.009
Sagittarius	35	7.493
Scorpius	10	12.480
Sculptor	86	3.159

Constellation	Rank	Overall brightness
Scutum	33	8.249
Serpens	67	5.652
Sextans	88	1.595
Taurus	12	12.292
Telescopium	49	6.759
Triangulum	27	9.101
Triangulum Australe	14	10.911
Tucana	74	5.092
Ursa Major	69	5.548
Ursa Minor	45	7.035
Vela	4	15.211
Virgo	79	4.481
Volans	19	9.904
Vulpecula	15	10.814

Possessive forms of the constellation names (with pronunciations)

Constellation	Possessive	Pronunciation
Andromeda	Andromedae	an drom′ uh die
Antlia	Antliae	ant′ lee eye
Apus	Apodis	ap′ oh diss
Aquarius	Aquarii	ah kwar′ ee ee
Aquila	Aquilae	ak′ will eye
Ara	Arae	air′ eye
Aries	Arietis	air ee′ ay tiss
Auriga	Aurigae	or eye′ guy
Boötes	Boötis	bow owe′ tiss
Caelum	Caeli	see′ lee
Camelopardalis	Camelopardalis	kam uh low par′ dah liss
Cancer	Cancri	kan′ kree
Canes Venatici	Canum Venaticorum	kay′ num ven at ih kor′ um
Canis Major	Canis Majoris	kay′ niss muh jor′ iss
Canis Minor	Canis Minoris	kay′ niss muh nor′ iss
Capricornus	Capricorni	kap rih corn′ ee
Carina	Carinae	kar ee′ nye
Cassiopeia	Cassiopeiae	kass ee oh pee′ eye
Centaurus	Centauri	sen tor′ ee
Cepheus	Cephei	see′ fee ee
Cetus	Ceti	set′ ee
Chamaeleon	Chamaeleontis	kuh meel ee on′ tiss
Circinus	Circini	sir sin′ ee
Columba	Columbae	kol um′ bye
Coma Berenices	Comae Berenices	koe′ my ber uh nye′ seez
Corona Australis	Coronae Australis	kor oh′ nye os tral′ iss
Corona Borealis	Coronae Borealis	kor oh′ nye bor ee al′ iss
Corvus	Corvi	kor′ vee
Crater	Crateris	kray′ ter iss
Crux	Crucis	kroo′ siss
Cygnus	Cygni	sig′ nee
Delphinus	Delphini	del fee′ nee
Dorado	Doradus	dor ah′ dus
Draco	Draconis	druh koe′ niss
Equuleus	Equulei	ek woo oo′ lay ee
Eridanus	Eridani	air uh day′ nee
Fornax	Fornacis	for nay′ siss
Gemini	Geminorum	jem uh nor′ um

Constellation	Possessive	Pronunciation
Grus	Gruis	groo′ eese
Hercules	Herculis	her′ kyoo liss
Horologium	Horologii	hor owe low′ gee ee
Hydra	Hydrae	hide′ rye
Hydrus	Hydri	hide′ ree
Indus	Indi	in′ dee
Lacerta	Lacertae	luh sir′ tie
Leo	Leonis	lee owe′ niss
Leo Minor	Leonis Minoris	lee owe′ niss my nor′ iss
Lepus	Leporis	lee por′ iss
Libra	Librae	lye′ bry
Lupus	Lupi	loo′ pee
Lynx	Lyncis	lin′ siss
Lyra	Lyrae	lie′ rye
Mensa	Mensae	men′ sigh
Microscopium	Microscopii	my krow skow′ pee ee
Monoceros	Monocerotis	mon awe sir awe′ tiss
Musca	Muscae	mus′ kye
Norma	Normae	nor′ mye
Octans	Octantis	ok tan′ tiss
Ophiuchus	Ophiuchi	off ee oo′ key
Orion	Orionis	or ee oh′ niss
Pavo	Pavonis	puh voe′ niss
Pegasus	Pegasi	peg′ uh see
Perseus	Persei	per′ see ee
Phoenix	Phoenicis	fen ee′ siss
Pictor	Pictoris	pik tor′ iss
Pisces	Piscium	pish′ ee um
Piscis Austrinus	Piscis Austrini	pie′ sis os tree′ nee
Puppis	Puppis	pup′ iss
Pyxis	Pyxidis	pik′ si diss
Reticulum	Reticuli	reh tik′ yoo lee
Sagitta	Sagittae	suh jeet′ eye
Sagittarius	Sagittarii	sa jit air′ ee ee
Scorpius	Scorpii	skor′ pee ee
Sculptor	Sculptoris	skulp tor′ iss
Scutum	Scuti	skoo′ tee
Serpens	Serpentis	sir pen′ tiss
Sextans	Sextantis	sex tan′ tiss
Taurus	Tauri	tor′ ee
Telescopium	Telescopii	tel es koe′ pee ee
Triangulum	Trianguli	try ang′ yoo lee
Triangulum Australe	Trianguli Australis	try ang′ yoo lee os tral′ iss
Tucana	Tucanae	too kan′ eye

Constellation	Possessive	Pronunciation
Ursa Major	Ursae Majoris	er' sigh muh jor' iss
Ursa Minor	Ursae Minoris	er' sigh muh nor' iss
Vela	Velorum	vee lor' um
Virgo	Virginis	ver' jin iss
Volans	Volantis	voe lan' tiss
Vulpecula	Vulpeculae	vul pek' yoo lye

Pronunciation key to constellation names

Andromeda	an draw′ meh duh
Antlia	ant′ lee ah
Apus	ape′ us
Aquarius	uh qwayr′ ee us
Aquila	ak′ will uh
Ara	air′ uh
Aries	air′ eeze
Auriga	or eye′ guh
Boötes	bow owe′ teez
Caelum	see′ lum
Camelopardalis	kam uh low par′ dah liss
Cancer	kan′ sir
Canes Venatici	kay′ neez ven ah tee′ see
Canis Major	kay′ niss may′ jor
Canis Minor	kay′ niss my′ nor
Capricornus	kap rih kor′ nus
Carina	kuh ree′ nuh
Cassiopeia	kass ee oh pee′ uh
Centaurus	sen tor′ us
Cepheus	see′ fee us
Cetus	see′ tus
Chamaeleon	kuh meel′ ee un
Circinus	sir sin′ us
Columba	kol um′ buh
Coma Berenices	koe′ muh bear uh nye′ seez
Corona Australis	kor oh′ nuh os tral′ iss
Corona Borealis	kor oh′ nuh boar ee al′ iss
Corvus	kor′ vus
Crater	kray′ ter
Crux	kruks
Cygnus	sig′ nus
Delphinus	dell fee′ nus
Dorado	dor ah′ doe
Draco	dray′ koe
Equuleus	ek woo oo′ lee us
Eridanus	air uh day′ nus
Fornax	for′ nax
Gemini	gem′ in eye
Grus	groose
Hercules	her′ cue leez
Horologium	hor uh low′ gee um
Hydra	hi′ druh

Hydrus	hi′ druss
Indus	in′ dus
Lacerta	luh sir′ tuh
Leo	lee′ owe
Leo Minor	lee′ owe my′ nor
Lepus	lee′ pus
Libra	lye′ bruh
Lupus	loo′ pus
Lynx	links
Lyra	lie′ ruh
Mensa	men′ suh
Microscopium	my krow scop′ ee um
Monoceros	mon oss′ sir us
Musca	mus′ kuh
Norma	nor′ muh
Octans	ok′ tans
Ophiuchus	off ee oo′ kus
Orion	or eye′ on
Pavo	pah′ voe
Pegasus	peg′ ah sus
Perseus	pur′ see us
Phoenix	fee′ niks
Pictor	pik′ tor
Pisces	pie′ seez
Piscis Austrinus	pie′ siss os try′ nus
Puppis	pup′ iss
Pyxis	pik′ siss
Reticulum	reh tik′ yoo lum
Sagitta	suh gee′ tuh
Sagittarius	sa ji tare′ ee us
Scorpius	skor′ pee us
Sculptor	skulp′ tor
Scutum	skoo′ tum
Serpens	sir′ pens
Sextans	sex′ tans
Taurus	tor′ us
Telescopium	tel es koe′ pee um
Triangulum	try ang′ yoo lum
Triangulum Australe	try ang′ yoo lum os trail′
Tucana	too kan′ uh
Ursa Major	er′ suh may′ jor
Ursa Minor	er′ suh my′ nor
Vela	vay′ luh
Virgo	ver′ go
Volans	voe′ lans
Vulpecula	vul pek′ yoo la

Proper motion – the 200 stars with the largest proper motion

Rank	Star	Constellation	Proper motion (seconds of arc per year)	Magnitude
1	Barnard's Star	Oph	10.29	9.5
2	Kapteyn's Star	Pic	8.70	10.0
3	Groombridge 1830	UMa	7.04	7.0
4	Lacaille 9352	PsA	6.9	8.6
5	Cordoba 32416	Scl	6.11	10.0
6	Ross 619	Cnc	5.40	14.2
7	61 Cyg A	Cyg	5.22	6.2
	61 Cyg B	Cyg	5.22	7.2
8	Lalande 21185	UMa	4.78	8.9
9	Wolf 359	Leo	4.71	15.7
10	ε Ind	Ind	4.69	5.9
11	Lalande 21258 A	UMa	4.53	10.2
	Lalande 21258 B	UMa	4.53	16.0
12	o^2 Eri A	Eri	4.08	5.3
	o^2 Eri B	Eri	4.08	9.8
	o^2 Eri C	Eri	4.08	12.3
13	Wolf 489	Vir	3.87	15.5
14	α Cen C	Cen	3.85	13.2
15	BD +5°1668	CMi	3.76	11.7
16	μ Cas	Cas	3.75	5.7
17	Wash. 5583	Lib	3.68	10.5
	Wash. 5584	Lib	3.68	9.8
18	α Cen A	Cen	3.68	0.8
	α Cen B	Cen	3.68	2.9
19	LP9–231	Dra	3.59	15.4
20	Cordoba 29191	Mic	3.46	7.9
21	UV Cet A	Cet	3.35	14.2
	UV Cet B	Cet	3.35	14.7
22	Luyten 789–6	Aqr	3.25	14.3
23	Ross 451	Dra	3.20	13.7
24	ε Eri	Eri	3.14	5.0
25	Ross 578	Eri	3.06	14.6
26	Van Maanen's Star	Psc	2.98	12.9
27	BD +66°717	UMa	2.96	10.6
28	Luyten 347–14	Sgr	2.93	13.7

Rank	Star	Constellation	Proper motion (seconds of arc per year)	Magnitude
29	BD +43°44A	And	2.89	9.5
	BD +43°44B	And	2.89	12.7
30	Luyten 192–72	Car	2.72	12.8
31	Luyten 145–141	Cen	2.68	12.5
32	BD +2°348	Cet	2.59	11.0
33	Luyten 720–88	Aqr	2.55	13.5
34	Ross 47	Ori	2.54	12.7
35	Luyten 230–188	Dor	2.53	15.0
36	Ross 695	Crv	2.52	12.4
37	Wolf 110	Psc	2.43	13.2
38	BD +6°398A	Cet	2.32	6.8
	BD +6°398B	Cet	2.32	13.3
39	Ross 769	Cap	2.30	12.4
40	BD +15°2620	Boo	2.30	9.9
41	BD +59°1915A	Dra	2.28	10.4
	BD +59°1915B	Dra	2.28	11.3
42	Arcturus	Boo	2.28	−0.0
43	β Hyi	Hyi	2.25	3.5
44	Luyten 768–119	Lib	2.24	12.6
45	BD −3°1123	Ori	2.24	9.1
46	BD +34°796	Per	2.20	9.3
47	Cordoba −51°532	Eri	2.20	7.0
48	Wolf 1106	Cep	2.14	14.4
49	BD +56°2966	Cas	2.09	6.4
50	Wolf 918	Aqr	2.06	12.2
51	ζ Tuc	Tuc	2.06	4.8
52	Luyten 97–12	Vol	2.05	15.0
53	Luyten 1159–16	Ari	2.04	13.7
54	Luyten 1274–3	Her	2.00	15.2
55	BD −20°4123	Lib	1.98	9.4
	BD −20°4125	Lib	1.98	7.0
56	BD +31°1684	Gem	1.97	8.8
57	BD +52°911	Cam	1.96	11.2
58	Wolf 851	Aql	1.96	13.3
59	BD −18°3019	Hya	1.94	12.9
60	τ Cet	Cet	1.92	4.1
61	Ross 791	Ari	1.91	13.7
62	BD +18°2776	Boo	1.90	10.9
63	Wolf 424	Vir	1.87	14.2
64	Wolf 1084	Cyg	1.87	16.8
65	σ Dra	Dra	1.84	5.5

Rank	Star	Constellation	Proper motion (seconds of arc per year)	Magnitude
66	Ross 248	And	1.82	13.8
67	Wolf 1062	Aql	1.80	12.7
68	BD −21°3781	Vir	1.80	9.7
69	Luyten 1912–9	UMa	1.80	13.0
70	Luyten 89–27	Hyi	1.79	13.5
71	BD −11°2741	Hya	1.79	11.4
72	Luyten 1115–12	Leo	1.76	15.0
73	BD +70°6810	Cas	1.76	10.7
74	BD +66°34A	Cas	1.76	11.8
	BD +66°34B	Cas	1.76	14.0
75	Luyten 1355–129	Vul	1.74	14.9
76	Luyten 513–23	For	1.73	14.6
77	Luyten 425–35	Gru	1.72	14.3
78	Luyten 205–128	Pav	1.72	12.9
79	BD +16°2658	Boo	1.71	12.3
80	Cordoba −46°293	Phe	1.71	11.8
81	Cordoba −32°5613	Pyx	1.69	11.8
82	Cordoba −33°4113	Pup	1.69	5.8
83	BD +53°1320	UMa	1.68	9.1
	BD +53°1321	UMa	1.68	9.2
84	Wolf 134	Ari	1.68	15.4
85	Wolf 1037	Peg	1.68	15.3
86	BD +0°3593	Oph	1.67	7.7
87	Luyten 362–29	Phe	1.66	14.5
88	Luyten 143–23	Car	1.65	15.3
89	Luyten 1499–28	Lyr	1.65	12.9
90	δ Pav	Pav	1.64	4.3
91	Cordoba −47°9919	Lup	1.64	6.2
92	BD +66°268	Cam	1.63	10.3
93	Cordoba −36°13940A	Sgr	1.63	6.5
	Cordoba −36°13940B	Sgr	1.63	13.0
94	Cordoba −57°6303	Nor	1.63	8.0
95	Wolf 1421	Gem	1.60	12.7
96	Cordoba −39°7301	Cen	1.59	5.5
97	Ross 34	Per	1.58	12.0
98	Luyten 750–52	Hya	1.58	14.3
99	BD +45°2505A	Her	1.58	10.7
	BD +45°2505B	Her	1.58	11.9
100	Wolf 33	Cas	1.58	12.5
101	Luyten 1346–53	Her	1.57	14.2
102	BD +63°137	Cas	1.55	10.2

Rank	Star	Constellation	Proper motion (seconds of arc per year)	Magnitude
103	Wolf 1056	And	1.54	11.7
104	Luyten 1545-14	LMi	1.54	14.0
105	Ross 322	And	1.52	14.7
106	BD −17°3813	Vir	1.52	5.4
107	BD +48°1829	UMa	1.51	11.8
108	Wolf 1130	Cyg	1.50	12.8
109	Luyten 879–14	Eri	1.49	13.9
110	Cordoba −65°253	Dor	1.49	10.1
111	Luyten 737–9	Lep	1.49	12.1
112	ζ^1 Ret	Ret	1.48	6.0
	ζ^2 Ret	Ret	1.48	5.7
113	Luyten 471–42	Cen	1.48	14.2
114	Wolf 611	Ser	1.48	14.8
	Wolf 612	Ser	1.48	16.1
115	Luyten 856–54A	Aqr	1.48	14.5
	Luyten 856–54B	Aqr	1.48	16.2
116	Luyten 1295–31	Peg	1.48	13.5
117	Luyten 440–30	For	1.47	13.3
118	BD +4°3195	Her	1.47	7.8
119	BD +53°2911AB	Lac	1.47	11.3
	BD +53°2911C	Lac	1.47	15.0
120	BD −4°4225	Oph	1.46	8.9
	BD −4°4226	Oph	1.46	11.3
121	BD +4°4048A	Aql	1.46	10.4
	BD +4°4048B	Aql	1.46	19.4
122	BD +50°1725	UMa	1.45	7.9
123	Ross 1027	CVn	1.45	13.7
124	ν Ind	Ind	1.45	5.9
125	Wolf 1057	Tau	1.44	14.0
126	Luyten 49–19	Oct	1.44	12.6
127	Luyten 750–42	Hya	1.43	14.7
128	Cordoba −38°1058	For	1.42	12.0
129	Luyten 1264–57	Com	1.42	14.5
130	Luyten 46–96	Oct	1.42	15.7
131	Ross 585	Per	1.41	14.1
132	Wolf 1039	Psc	1.41	12.4
133	BD +71°482A	UMa	1.40	10.1
	BD +71°482B	UMa	1.40	10.2
134	Ross 53	Boo	1.40	12.0
135	Luyten 791–76	Aqr	1.40	14.4

Rank	Star	Constellation	Proper motion (seconds of arc per year)	Magnitude
136	Luyten 578–71	Scl	1.39	14.7
	Luyten 578–72	Scl	1.39	15.2
137	Cordoba –25°225	Cet	1.38	6.7
138	BD +33°529	Per	1.38	10.8
139	Ross 128	Vir	1.38	12.6
140	BD +24°2733A	Boo	1.38	10.8
	BD +24°2733B	Boo	1.38	11.1
141	BD +4°123	Psc	1.37	6.7
142	BD +41°750A	Per	1.37	9.7
	BD +41°750B	Per	1.37	10.4
143	BD –0°2944	Lib	1.37	7.7
144	BD +22°3908	Vul	1.37	8.0
145	BD +1°4774	Psc	1.37	10.4
146	Luyten 868–3	Cet	1.36	14.6
147	Wolf 1324	Per	1.36	11.8
148	Wolf 534	Vir	1.36	14.5
149	Cordoba –31°6229	Pyx	1.35	7.1
150	Luyten 1278–24	Her	1.35	11.2
151	Ross 165	Vul	1.34	13.8
152	Luyten 725–32	Cet	1.33	13.1
153	Cordoba –42°469	Phe	1.33	11.5
154	γ Ser	Ser	1.33	4.3
155	BD +2°3312	Oph	1.33	8.9
156	Sirius	CMa	1.32	–1.5
157	BD +68°946	Dra	1.31	10.7
158	Cordoba –28°302	Scl	1.30	12.5
159	Cordoba –70°340	Dor	1.30	8.8
160	Greenwich +82 1111	Cam	1.30	11.6
161	BD +26°4734	Peg	1.30	6.3
162	BD –18°359	Cet	1.29	11.6
163	BD –0°4470	Psc	1.29	10.9
164	BD –14°6437	Aqr	1.29	8.8
165	Luyten 1190–34	Vir	1.28	15.5
166	BD +18°683	Tau	1.27	11.2
167	Luyten 815–20	CMa	1.27	15.5
168	Luyten 535–3	Ant	1.27	12.8
169	Luyten 210–70	Tel	1.27	14.2
170	ι Per	Per	1.26	4.6
171	Luyten 745–46A	Pup	1.26	12.9
	Luyten 745–46B	Pup	1.26	17.6

Rank	Star	Constellation	Proper motion (seconds of arc per year)	Magnitude
172	Luyten 395–13	Cen	1.26	13.8
173	Cordoba –26°8883	Hya	1.26	8.3
174	Wolf 1465	Sct	1.26	15.3
175	Cordoba –27°14659	Cap	1.26	6.7
176	Wolf 219	Tau	1.25	15.1
177	Procyon	CMi	1.25	0.9
178	Cordoba –80°328	Cha	1.25	11.0
179	BD +11°2576	Vir	1.25	10.5
180	BD –7°3856	Lib	1.25	11.2
181	BD –7°4003	Lib	1.25	12.3
182	Greenwich +76 5308	UMa	1.25	13.0
183	Luyten 205–83	Ara	1.25	13.6
184	Luyten 22–69	Oct	1.25	14.4
185	Luyten 1545–74	LMi	1.24	15.0
186	Luyten 194–11	Cru	1.24	15.2
187	Luyten 1194–26	Vir	1.24	13.5
188	Ross 508	Ser	1.24	15.4
189	Cordoba –26°12026A	Oph	1.24	6.4
	Cordoba –26°12026B	Oph	1.24	6.4
	Cordoba –26°12036	Oph	1.24	7.7
190	BD +5°3993	Ser	1.24	10.7
191	Luyten 1945–4	Cam	1.23	14.0
192	Luyten 1707–1	Her	1.23	11.7
193	Wolf 1040	Psc	1.23	14.2
194	η Cas A	Cas	1.22	4.0
	η Cas B	Cas	1.22	8.7
195	BD –5°1123	Eri	1.22	7.2
196	Wolf 397	Leo	1.22	11.4
197	Cordoba –25°10553A	Hya	1.22	13.1
	Cordoba –25°10553B	Hya	1.22	13.2
198	Cordoba –37°10765A	Sco	1.22	12.0
	Cordoba –37°10765B	Sco	1.22	16.0
199	Luyten 1293–88	Peg	1.22	13.2
200	Luyten 455–111	Pup	1.21	13.4

The size of the constellations

1 BY TOTAL AREA

Rank	Name	Area (sq. deg)	% of sky
1	Hydra	1302.84	3.158
2	Virgo	1294.43	3.138
3	Ursa Major	1279.66	3.102
4	Cetus	1231.41	2.985
5	Hercules	1225.15	2.970
6	Eridanus	1137.92	2.758
7	Pegasus	1120.79	2.717
8	Draco	1082.95	2.625
9	Centaurus	1060.42	2.571
10	Aquarius	979.85	2.375
11	Ophiuchus	948.34	2.299
12	Leo	946.96	2.296
13	Boötes	906.83	2.198
14	Pisces	889.42	2.156
15	Sagittarius	867.43	2.103
16	Cygnus	803.98	1.949
17	Taurus	797.25	1.933
18	Camelopardalis	756.83	1.835
19	Andromeda	722.28	1.751
20	Puppis	673.43	1.633
21	Auriga	657.44	1.594
22	Aquila	652.47	1.582
23	Serpens	636.92	1.544
24	Perseus	615.00	1.491
25	Cassiopeia	598.41	1.451
26	Orion	594.12	1.440
27	Cepheus	587.79	1.425
28	Lynx	545.39	1.322
29	Libra	538.05	1.304
30	Gemini	513.76	1.245
31	Cancer	505.87	1.226
32	Vela	499.65	1.211
33	Scorpius	496.78	1.204
34	Carina	494.18	1.198
35	Monoceros	481.57	1.167
36	Sculptor	474.76	1.151
37	Phoenix	469.32	1.138

Rank	Name	Area (sq. deg)	% of sky
38	Canes Venatici	465.19	1.128
39	Aries	441.39	1.070
40	Capricornus	413.95	1.003
41	Fornax	397.50	0.964
42	Coma Berenices	386.47	0.937
43	Canis Major	380.11	0.921
44	Pavo	377.67	0.916
45	Grus	365.51	0.886
46	Lupus	333.68	0.809
47	Sextans	313.51	0.760
48	Tucana	294.56	0.714
49	Indus	294.01	0.713
50	Octans	291.05	0.706
51	Lepus	290.29	0.704
52	Lyra	286.48	0.694
53	Crater	282.40	0.685
54	Columba	270.18	0.655
55	Vulpecula	268.17	0.650
56	Ursa Minor	255.86	0.620
57	Telescopium	251.51	0.610
58	Horologium	248.88	0.603
59	Pictor	246.73	0.598
60	Piscis Austrinis	245.37	0.595
61	Hydrus	243.04	0.589
62	Antlia	238.90	0.579
63	Ara	237.06	0.575
64	Leo Minor	231.96	0.562
65	Pyxis	220.83	0.535
66	Microscopium	209.51	0.508
67	Apus	206.32	0.500
68	Lacerta	200.69	0.487
69	Delphinus	188.54	0.457
70	Corvus	183.80	0.446
71	Canis Minor	183.37	0.445
72	Dorado	179.17	0.434
73	Corona Borealis	178.71	0.433
74	Norma	165.29	0.401
75	Mensa	153.48	0.372
76	Volans	141.35	0.343
77	Musca	138.36	0.335
78	Triangulum	131.85	0.320
79	Chamaeleon	131.59	0.319
80	Corona Australis	127.69	0.310

Rank	Name	Area (sq. deg)	% of sky
81	Caelum	124.86	0.303
82	Reticulum	113.94	0.276
83	Triangulum Australe	109.98	0.267
84	Scutum	109.11	0.265
85	Circinus	93.35	0.226
86	Sagitta	79.93	0.194
87	Equuleus	71.64	0.174
88	Crux	68.45	0.166

2 IN ALPHABETICAL ORDER

Constellation	Rank	Area (sq. deg)
Andromeda	19	722.28
Antlia	62	238.90
Apus	67	206.32
Aquarius	10	979.85
Aquila	22	652.47
Ara	63	237.06
Aries	39	441.39
Auriga	21	657.44
Boötes	13	906.83
Caelum	81	124.86
Camelopardalis	18	756.83
Cancer	31	505.87
Canes Venatici	38	465.19
Canis Major	43	380.11
Canis Minor	71	183.37
Capricornus	40	413.95
Carina	34	494.18
Cassiopeia	25	598.41
Centaurus	9	1060.42
Cepheus	27	587.79
Cetus	4	1231.41
Chamaeleon	79	131.59
Circinus	85	93.35
Columba	54	270.18
Coma Berenices	42	386.47
Corona Australis	80	127.69

Constellation	Rank	Area (sq. deg)
Corona Borealis	73	178.71
Corvus	70	183.80
Crater	53	282.40
Crux	88	68.45
Cygnus	16	803.98
Delphinus	69	188.54
Dorado	72	179.17
Draco	8	1082.95
Equuleus	87	71.64
Eridanus	6	1137.92
Fornax	41	397.50
Gemini	30	513.76
Grus	45	365.51
Hercules	5	1225.15
Horologium	58	248.88
Hydra	1	1302.84
Hydrus	61	243.04
Indus	49	294.01
Lacerta	68	200.69
Leo	12	946.96
Leo Minor	64	231.96
Lepus	51	290.29
Libra	29	538.05
Lupus	46	333.68
Lynx	28	545.39
Lyra	52	286.48
Mensa	75	153.48
Microscopium	66	209.51
Monoceros	35	481.57
Musca	77	138.36
Norma	74	165.29
Octans	50	291.05
Ophiuchus	11	948.34
Orion	26	594.12
Pavo	44	377.67
Pegasus	7	1120.79
Perseus	24	615.00
Phoenix	37	469.32
Pictor	59	246.73
Pisces	14	889.42
Piscis Austrinus	60	245.37
Puppis	20	673.43
Pyxis	65	220.83

Constellation	Rank	Area (sq. deg)
Reticulum	82	113.94
Sagitta	86	79.93
Sagittarius	15	867.43
Scorpius	33	496.78
Sculptor	36	474.76
Scutum	84	109.11
Serpens	23	636.92
Sextans	47	313.51
Taurus	17	797.25
Telescopium	57	251.51
Triangulum	78	131.85
Triangulum Australe	83	109.98
Tucana	48	294.56
Ursa Major	3	1279.66
Ursa Minor	56	255.86
Vela	32	499.65
Virgo	2	1294.43
Volans	76	141.35
Vulpecula	55	268.17

Solar conjunction dates for the constellations

1 BY DATE

3 Jan	Lyra	17 Jun	Pictor
6 Jan	Sagittarius	18 Jun	Columba
10 Jan	Telescopium	22 Jun	Auriga
14 Jan	Pavo	4 Jul	Canis Major
15 Jan	Aquila	8 Jul	Gemini
15 Jan	Sagitta	8 Jul	Monoceros
24 Jan	Vulpecula	11 Jul	Puppis
30 Jan	Cygnus	16 Jul	Canis Minor
31 Jan	Delphinus	19 Jul	Volans
4 Feb	Microscopium	22 Jul	Lynx
5 Feb	Capricornus	1 Aug	Cancer
7 Feb	Equuleus	2 Aug	Carina
19 Feb	Indus	4 Aug	Camelopardalis
24 Feb	Aquarius	6 Aug	Pyxis
24 Feb	Piscis Austrinus	18 Aug	Vela
27 Feb	Grus	25 Aug	Leo Minor
27 Feb	Lacerta	26 Aug	Antlia
2 Mar	Pegasus	26 Aug	Sextans
18 Mar	Tucana	31 Aug	Leo
29 Mar	Pisces	1 Sep	Chamaeleon
29 Mar	Sculptor	10 Sep	Ursa Major
3 Apr	Andromeda	11 Sep	Crater
5 Apr	Phoenix	15 Sep	Hydra
11 Apr	Cassiopeia	21 Sep	Ursa Minor
17 Apr	Cetus	27 Sep	Corvus
24 Apr	Triangulum	27 Sep	Crux
26 Apr	Cepheus	29 Sep	Musca
26 Apr	Hydrus	3 Oct	Coma Berenices
31 Apr	Aries	7 Oct	Centaurus
4 May	Fornax	8 Oct	Canes Venatici
9 May	Perseus	12 Oct	Virgo
11 May	Eridanus	29 Oct	Circinus
11 May	Horologium	1 Nov	Boötes
21 May	Reticulum	8 Nov	Draco
2 Jun	Caelum	8 Nov	Libra
2 Jun	Taurus	8 Nov	Lupus
10 Jun	Dorado	18 Nov	Corona Borealis
14 Jun	Mensa	21 Nov	Norma
15 Jun	Lepus	21 Nov	Triangulum Australe
15 Jun	Orion	22 Nov	Apus

4 Dec	Scorpius	12 Dec	Ophiuchus
5 Dec	Serpens	31 Dec	Corona Australis
9 Dec	Ara	31 Dec	Scutum
12 Dec	Hercules	—	Octans

2 IN ALPHABETICAL ORDER

Andromeda	3 Apr	Hercules	12 Dec
Antlia	26 Aug	Horologium	11 May
Apus	22 Nov	Hydra	15 Sep
Aquarius	24 Feb	Hydrus	26 Apr
Aquila	15 Jan	Indus	19 Feb
Ara	9 Dec	Lacerta	27 Feb
Aries	31 Apr	Leo	31 Aug
Auriga	22 Jun	Leo Minor	25 Aug
Boötes	1 Nov	Lepus	15 Jun
Caelum	2 Jun	Libra	8 Nov
Camelopardalis	4 Aug	Lupus	8 Nov
Cancer	1 Aug	Lynx	22 Jul
Canes Venatici	8 Oct	Lyra	3 Jan
Canis Major	4 Jul	Mensa	14 Jun
Canis Minor	16 Jul	Microscopium	4 Feb
Capricornus	5 Feb	Monoceros	8 Jul
Carina	2 Aug	Musca	29 Sep
Cassiopeia	11 Apr	Norma	21 Nov
Centaurus	7 Oct	Octans	—
Cepheus	26 Apr	Ophiuchus	12 Dec
Cetus	17 Apr	Orion	15 Jun
Chamaeleon	1 Sep	Pavo	14 Jan
Circinus	29 Oct	Pegasus	2 Mar
Columba	18 Jun	Perseus	9 May
Coma Berenices	3 Oct	Phoenix	5 Apr
Corona Australis	31 Dec	Pictor	17 Jun
Corona Borealis	18 Nov	Pisces	29 Mar
Corvus	27 Sep	Piscis Austrinus	24 Feb
Crater	11 Sep	Puppis	11 Jul
Crux	27 Sep	Pyxis	6 Aug
Cygnus	30 Jan	Reticulum	21 May
Delphinus	31 Jan	Sagitta	15 Jan
Dorado	10 Jun	Sagittarius	6 Jan
Draco	8 Nov	Scorpius	4 Dec
Equuleus	7 Feb	Sculptor	29 Mar
Eridanus	11 May	Scutum	31 Dec
Fornax	4 May	Serpens	5 Dec
Gemini	8 Jul	Sextans	26 Aug
Grus	27 Feb	Taurus	2 Jun

Telescopium	10 Jan	Ursa Minor	21 Sep
Triangulum	24 Apr	Vela	18 Aug
Triangulum Australe	21 Nov	Virgo	12 Oct
Tucana	18 Mar	Volans	19 Jul
Ursa Major	10 Sep	Vulpecula	24 Jan

Star designations

To designate the 24 brightest stars in a constellation, Johannes Bayer (1572–1625) used the lower case letters of the Greek alphabet:

Letter	Lower Case	Upper Case
alpha	α	A
beta	β	B
gamma	γ	Γ
delta	δ	Δ
epsilon	ε	E
zeta	ζ	Z
eta	η	H
theta	θ	Θ
iota	ι	I
kappa	κ	K
lambda	λ	Λ
mu	μ	M
nu	ν	N
xi	ξ	Ξ
omicron	o	O
pi	π	Π
rho	ρ	P
sigma	σ	Σ
tau	τ	T
upsilon	υ	Y
phi	φ	Φ
chi	χ	X
psi	φ	Ψ
omega	ω	Ω

for other stars:

When the 24 letters of the Greek alphabet were used up, Bayer employed lower case Roman letters (a, b, c. . .). When these 26 letters were exhausted, he assigned upper case Roman letters (A, B, C. . .).

Fainter stars are designated in a variety of ways. If the star is visible to the naked-eye, but has no Bayer designation, often the star is referred to by its Flamsteed number. John Flamsteed (1646–1719) compiled a star catalogue which was published in 1725, six years after his death. In this catalogue, called the *Historia Coelestis Britannica*, Flamsteed assigned numbers to the stars within each constellation according to the stars' right ascensions. Thus the star 31 Lyn is the 31st star within the constellation Lynx, according to Flamsteed's catalog.

Hundreds of thousands of even fainter stars are listed in many catalogs. The most popular are the SAO (*Smithsonian Astrophysical Observatory Catalogue*, 1966 with 258 997 stars), the BD (*Bonner Durchmusterung*, 1862 with 324 189 stars originally, but extended by 1930 to 1 072 000 stars), and the HD (*Henry Draper Catalogue*, 1924 with spectral types for 225 300 stars).

In 1862, Friedrich Wilhelm August Argelander (1799–1875) assigned the upper case Roman letters R–Z to variable stars within a constellation. After these nine letters came double upper case letters (RR–RZ, SS–SZ, AA–AZ, BB–BZ, QQ-QZ). Since no 'J' was used to begin these pairs, an additional 334 stars could be denoted. Thus, R CrB is the first variable star in the constellation of Corona Borealis and QZ Ori is the 343rd variable star in Orion. If even more are needed, the single upper case letter 'V' is used. Therefore, Plaskett's Star is designated as V640 Mon, the 640th variable star in the constellation of Monoceros.

Novae are designated initially by constellation and year, as for example, Nova Cygni 1975. When the brightness of the event diminishes, such objects are assigned a variable star designation. Thus, the above object is now known as V1500 Cyg.

Star names

Star name	Designation	Visual magnitude
Acamar	θ Eri	2.9
Achernar	α Eri	0.5
Achird	η Cas	3.6
Acrab	β Sco	2.6
Acrux	α Cru	0.8
Acubens	α Cnc	4.3
Adara	ε CMa	1.5
Adhafera	ζ Leo	3.6
Adhara	ε CMa	1.5
Adhil	ν And	4.2
Agena	β Cen	0.6
Ain	ε Tau	3.6
Ain al Rami	μ¹ Sgr	4.8
Ak	α UMa	1.8
Aladfar	η Lyr	4.4
Aladfar	μ Lyr	5.1
Alamak	γ And	2.0
Al Anchat al Nahr	τ² Eri	4.8
Alanf (Al Anf)	ε Peg	2.4
Alanz (Al Anz)	ε Aur	3.0
Alaraph	α Vir	1.0
Alaraph	β Vir	3.6
Alaraph	ε Vir	2.8
Alascha	λ Sco	1.6
Alathfar (Al Athfar)	μ Lyr	5.1
Al Atik	o Per	3.8
Albaldah (Al Baldah)	π Sgr	2.9
Albali (Al Bali)	ε Aqr	3.8
Albireo (Albereo)	β Cyg	3.0
Alchiba (Al Chiba, Alkhiba, Alchita)	α Crv	4.0
Alcor	80 UMa	4.0
Alcyone	ζ Tau	3.0
Aldebaran	α Tau	0.9
Alderamin (Alderaimin)	α Cep	2.4
Aldhafara (Aldhafera)	ζ Leo	3.4
Al Dhanab	γ Gru	3.0
Al Dhiba	ι Dra	3.3
Aldhibah (Al Dibah)	ζ Dra	3.2
Al Dhihi	ι Dra	3.3
Aldib	δ Dra	3.1

Star name	Designation	Visual magnitude
Alfard	α Hya	2.0
Alfecca	α CrA	4.1
Alfecca Meridiana	α CrA	4.1
Alfirk	β Cep	3.3
Alga	θ Ser	4.5
Algebar	β Ori	0.3
Algedi (Algiedi)	'α' Cap[a]	
Algedi Prima	α¹ Cap	4.2
Algedi Secunda	α² Cap	3.6
Algeiba	γ Leo	1.9
Algenib	γ Peg	2.8
Algenib	α Per	1.8
Algenubi	ε Leo	3.0
Algieba (Al Gieba)	γ Leo	1.9
Algol	β Per	2.1
Algomeyla	β CMi	2.9
Algomeysa	α CMi	0.4
Algorab (Algoral, Algorel, Algores)	δ Crv	3.0
Alhajoth	α Aur	0.2
Al Hammam	ζ Peg	3.4
Alhena	γ Gem	1.9
Alioth (Aliath)	ε UMa	1.8
Al Kaff al Jidmah	γ Cet	3.5
Alkaid	η UMa	1.9
Al Kalb al Asad	α Leo	1.3
Al Kalb al Rai	ρ Cep	5.5
Alkalurops	μ Boo	4.5
Alkaphrah (Al Kaphrah)	χ UMa	3.7
Alkes	α Crt	4.2
Alkhiba	α Crv	4.0
Al Kirduh (Alkirdah, Alkurhah)	ξ Cep	4.3
Almaac (Almak, Almaak, Almaack)	γ And	2.2
Almaaz	ε Aur	3.0v[b]
Almach (Almaach)	γ And	2.2
Al Mankib	α Ori	0.5
Almeisan	γ Gem	1.9
Al Minliar al Asad	κ Leo	4.5
Al Minliar al Ghurab	α Crv	4.0
Al Minliar al Shuja	σ Hya	4.4
Al Mizar	β And	2.1
Almuredin	ε Vir	2.8
Alnair (Al Nair, Al Na'ir)	α Gru	1.7
Al Nasl (Alnasl)	γ Sgr	3.0
Alnath	α Ari	2.0

Star name	Designation	Visual magnitude
Alnath	β Tau	1.6
Alnilam (Alnihan, Alnitam)	ε Ori	1.7
Alnitak (Alnitah)	ζ Ori	1.8
Al Niyat (Alniyat)	σ Sco	2.9
Al Niyat	τ Sco	2.8
Alphard (Alphart)	α Hya	2.0
Alphecca (Alphacca, Alphekka, Alphaca)	α CrB	2.2
Alpheratz (Alpherat)	α And	2.1
Alphirk	β Cep	3.2
Alrai	γ Cep	3.9
Al Rakis	ν Dra	5.0
Alrami	α Sgr	4.1
Alrischa (Al Rischa, Alrisha, Al Rescha, Alrescha, Al Richa)	α Psc	4.3
Alruccabah	α UMi	2.0
Al Rukbah al Dajajah	ω² Cyg	5.4
Alsafi	σ Dra	4.7
Alsahm	α Sge	4.4
Al Sanam al Nakah	β Cas	2.3
Alsciaukat	31 Lyn	4.3
Alshain (Alschain, Alschairn)	β Aql	3.7
Alshat	ν Cap	4.6
Alshemali	μ Leo	3.9
Al Sheratain	β Ari	2.6
Alsuhail (Al Suhail al Wazn)	λ Vel	2.2
Al Suhail al Muhlif	γ Vel	1.8
Altair	α Aql	0.8
Altais	δ Dra	3.2
Altarf (A! Tarf)	β Cnc	3.5
Alterf	λ Leo	4.3
Althafi	σ Dra	4.7
Al Tinnin	α Dra	3.7
Aludra	η CMa	2.5
Alula Austrais	ξ UMa	3.9
Alula Borealis	ν UMa	3.7
Alwaid	β Dra	2.8
Alwazl	γ Sgr	2.9
Alwazn (Al Wazor)	δ CMa	1.8
Alya	θ Ser	4.5
Alzirr	ξ Gem	3.4
Amazon Star	γ Ori	1.7
Ancha	θ Aqr	4.3
Anchat	τ² Eri	4.8

Star name	Designation	Visual magnitude
Angel Stern	α UMi	2.0
Angetenar	τ² Eri	4.8
Ankaa	α Phe	2.4
Anser	α Vul	4.4
Antares	α Sco	1.0
Antecanis	α CMi	0.4
Apollo	α Gem	1.6
Arcturus	α Boo	0.0
Arich	γ Vir	2.9
Arided (Aridif)	α Cyg	1.3
Arietis	α Ari	2.2
Arkab	β¹ Sgr	4.3
Arkeb Posterior	β² Sgr	4.5
Arkeb Prior	β¹ Sgr	4.3
Arneb	α Lep	2.6
Arrai	γ Cep	3.2
Arrakis	μ Dra	5.1
Arrioph	α Cyg	1.3
Ascella	ζ Sgr	2.6
Aschere	α CMa	−1.5
Asellus Australis	δ Cnc	4.2
Asellus Boraelis	γ Cnc	4.7
Asellus Primus	θ Boo	4.1
Asellus Secundus	ι Boo	4.8
Asellus Tertius	κ Boo	4.5
Ashtaroth	α CrB	2.2
Asmidiske (Aspidiske, Azmidiske)	ξ Pup	3.5
Asmidiske (Aspidiske, Azmidiske)	ι Car	2.2
Asterion	β CVn	4.3
Asterope	21 Tau	5.8
Asuia	β Dra	2.8
Atair	α Aql	0.9
Athafi (Athafiyy)	σ Dra	4.7
Atik (Ati, Atiks)	o Per	3.9
Atlas	27 Tau	3.8
Atria	α TrA	1.9
Auva	δ Vir	3.4
Avior	ε Car	1.9
Azelfafage	π¹ Cyg	4.8
Azha	η Eri	4.1
Azimech	α Vir	1.2
Baham	θ Peg	3.5
Baten Kaitos (Batenkaitos)	ζ Cet	3.9
Becrux	β Cru	1.3

Star name	Designation	Visual magnitude
Beid	o¹ Eri	4.1
Bellatrix	γ Ori	1.6
Benetnash (Benatnasch, Benetnasch)	η UMa	1.9
Betelgeuse (Beteigeux, Betelgeuze)	α Ori	0.5
Biham	θ Peg	3.5
Botein	δ Ari	4.5
Brachium	σ Lib	3.3
Bunda	ξ Aqr	4.7
Caiam (Cajam)	ω Her	4.6
Calbalakrab	α Sco	1.0
Calx	μ Gem	2.9
Canicula	α CMa	−1.6
Canopus	α Car	−0.7
Capella	α Aur	0.1
Caph	β Cas	2.3
Caput Trianguli	α Tri	3.4
Castor	α Gem	1.6
Castula	ν¹ Cas	4.8
Castula	ν² Cas	4.6
Cebalrai (Cebelrai, Celb-al-Rai)	β Oph	2.8
Ceginus	γ Boo	3.0
Celaeno (Celeno, Celieno)	16 Tau	5.4
Chaph	β Cas	2.4
Chara	β CVn	4.3
Cheleb	β Oph	2.8
Chertan	θ Leo	3.4
Chort	θ Leo	3.4
Cih	γ Cas	2.5v[b]
Clava	μ Boo	4.3
Cor Caroli	α CVn	2.9
Cor Hydrae	α Hya	2.0
Cor Leonis	α Leo	1.4
Cornu	σ Lib	3.3
Cor Scorpii	α Sco	1.0
Cor Serpentis	α Ser	2.7
Cor Tauri	α Tau	0.9
Coxa	θ Leo	3.4
Cujam	ω Her	4.6
Cursa	β Eri	2.9
Cymbae	α Phe	2.4
Cynosaura	α UMi	2.0
Dabih	'β' Cap[a]	
Dabih Major	β¹ Cap	3.1
Dabih Minor	β² Cap	6.1

Star name	Designation	Visual magnitude
Deneb	α Cyg	1.3
Deneb Aleet (Deneb)	β Leo	2.1
Deneb Algedi (Deneb Algiedi)	δ Cap	2.9
Deneb Algenubi (Deneb)	η Cet	3.5
Deneb al Schemali	ι Cet	3.6
Deneb Cygni (Deneb)	α Cyg	1.3
Deneb Dulfim (Deneb)	ε Del	4.0
Deneb el Adige	α Cyg	1.3
Deneb el Delphinus	ε Del	4.0
Deneb el Okab (Deneb)	ε Aql	4.0
Deneb el Okab (Deneb)	ζ Aql	3.0
Deneb Kaitos (Deneb Kaitos Senubi)	β Cet	2.0
Deneb Kaitos Shemali (Deneb Kaitos Shamaliyy)	ι Cet	3.6
Deneb Okab	δ Aql	3.4
Denebola	β Leo	2.1
Dhabih	β Cap	3.1
Dhalim	β Eri	2.8
Dheneb	η Cet	3.5
Dhur (Duhr)	δ Leo	2.6
Diadem	α Com	5.2
Difda al Auwel	α PsA	1.2
Diphda (Difda, Difda al Thani)	β Cet	2.2
Dnoces	ι UMa	3.1
Dschubba	δ Sco	2.3
Dubhe (Dubb)	α UMa	1.8
Dziban (Dsiban)	ψ Dra	4.9
Ed Asich (Edasich, Eldsich)	ι Dra	3.5
El Acola	ξ UMa	3.8
Elacrab	β Sco	2.6
El Dhalim	β Eri	2.8
El Difda	β Cet	2.0
Electra	17 Tau	2.8
Elgebar	β Ori	0.1
El Ghoul	β Per	2.1
Elgomaisa	α CMi	0.4
El Kaprah	κ UMa	3.6
El Karidab	δ Sgr	2.7
Elkeid	η UMa	1.9
El Khereb	τ Peg	4.6
Elkhiffa Australis	α Lib	2.7
Elkhiffa Boraelis	β Lib	2.6
El Koprah	χ UMa	3.7
Elmathalleth (Elmuthalleth)	α Tri	3.4

Star name	Designation	Visual magnitude
El Melik	α Aqr	3.0
El Nath (Elnath)	β Tau	1.7
El Nath	α Ari	2.0
El Phekrah	μ UMa	3.1
El Rischa	α Psc	4.3
Eltanin	γ Dra	2.2
Enif (Enf, Eniph, Enir)	ε Peg	2.4
Erakis	μ Cep	4.1v[b]
Er Rai (Errai)	γ Cep	3.2
Errakis (Er Rakis, El Rakis)	μ Dra	4.9
Etamin (Etanin, Ettanin)	γ Dra	2.4
Falx Italica	38 Boo	5.7
Fidis	α Lyr	0.0
Fom	ε Peg	2.5
Fomalhaut	α PsA	1.2
Fornacis	α For	3.9
Fum Al Samakah	β Psc	4.5
Furud	ζ CMa	3.0
Gacrux	γ Cru	1.6
Gallina	α Cyg	1.3
Garnet Star	μ Cep	4.1
Gemma	α CrB	2.3
Genam	ξ Dra	3.8
Giansar (Gianfar, Giausar, Giauzar)	λ Dra	4.1
Giedi	'α' Cap[a]	
Giedi Prima	α¹ Cap	4.2
Giedi Secunda	α² Cap	3.6
Gienah (Gienah Cygni)	ε Cyg	2.5
Gienah (Gienah Ghurab)	γ Crv	2.6
Gildun	δ UMi	4.4
Gnosia (Gnosia Stella Coronae)	α CrB	2.2
Gomeisa (Gomelza)	β CMi	2.9
Gorgona	β Per	2.1
Gorgonea Prima	β Per	2.1
Gorgonea Quarta	ω Per	4.6
Gorgonea Secunda	π Per	4.7
Gorgonea Tertia	ρ Per	3.4
Graffias (Grafias)	ζ Sco	3.8
Gredi	α² Cap	3.6
Grumium	ξ Dra	3.9
Hadar	β Cen	0.6
Hamal (Hamul, Hemal)	α Ari	2.0
Haris	γ Boo	3.0
Hasseleh	ι Aur	2.7

Star name	Designation	Visual magnitude
Hastorang	α PsA	1.3
Hatsya (Hatysa)	ι Ori	2.8
Heka	λ Ori	3.7
Hercules	β Gem	1.1
Herschel's Garnet Star	μ Cep	4.1
Heze	ζ Vir	3.4
Hoedus I (Haedus)	ζ Aur	3.8
Hoedus II	η Aur	3.2
Homam (Homan, Humam)	ζ Peg	3.6
Hyadem I (Hyadum I, Hyadem Primus)	γ Tau	3.7
Hyadem II (Hyadum II)	δ¹ Tau	3.8
Icalurus (Inkalunis)	μ Boo	4.3
Iclarkrau (Iclarkrav)	δ Sco	2.3
Isis	α CMa	−1.5
Isis	γ CMa	4.1
Izar	ε Boo	2.7
Jabbah	ν Sco	4.0
Jed	δ Oph	2.7
Jewel	α CrB	2.2
Job's Star	α Boo	−0.04
Jugum	γ Lyr	3.2
Juza	λ Dra	3.8
Kabeleced	α Leo	1.4
Kaff	β Cas	2.3
Kaffa	δ UMa	3.4
Kaffaljidhmah	γ Cet	3.6
Kaitain	α Psc	4.3
Kajam	ι Her	4.5
Kalb (Kelb)	α Leo	1.4
Kalb al Akrab	α Sco	1.0
Kalb al Rai (Kalbalrai)	β Oph	2.8
Kalbelaphard	α Hya	2.0
Kaus Australis	ε Sgr	1.9
Kaus Borealis	λ Sgr	2.8
Kaus Meridionalis (Kaus Media)	δ Sgr	2.7
Keid (Kied)	o² Eri	4.4
Kelb Alrai (Kelb-al-Rai)	β Oph	2.9
Kerb	τ Peg	4.6
Kiffa Australis	α Lib	2.9
Kiffa Boraelis	β Lib	2.7
Kitalpha (Kitabpha, Kitalphar, Kitel Phard)	α Equ	3.9
Kochab (Kocab, Kochah)	β UMi	2.1

Star name	Designation	Visual magnitude
Kornephoros (Korneforos)	β Her	2.8
Kraz	β Crv	2.7
Ksora	δ Cas	2.8
Kuma	ν Dra	5.0
Kurhah	ξ Cep	4.3
Kursa	β Eri	2.8
La Superba	y CVn	5.6
Lesath	v Sco	4.3
Lesath (Lesuth, Leschath)	υ Sco	2.7
Lesath (Lesuth)	λ Sco	1.6
Lodestar	α UMi	2.0
Lucida Cymbae	α Phe	2.4
Maasym (Masym)	λ Her	4.5
Maaz	ε Aur	3.0
Mabsuthat	31 Lyn	4.3
Maia	20 Tau	4.0
Marfak (Marfac)	α Per	1.8
Marfak	θ Cas	4.3
Marfak	μ Cas	5.2
Marfak (Marfik)	κ Her	5.3
Marfik (Marfic, Marsic)	λ Oph	3.8
Markab (Marchab)	α Peg	2.5
Markab (Markeb)	τ Peg	4.6
Markeb (Markab)	k¹ Pup	4.5
Markeb	x Vel	3.4
Marrha	38 Boo	5.7
Marsik	κ Her	5.3
Matar	η Peg	2.9
Mebsuta (Mebusta)	ε Gem	3.0
Media	δ Sgr	2.7
Megrez (Megres)	δ UMa	3.4
Meissa	λ Ori	3.4
Mekbuda	ζ Gem	3.8
Melboula	ε Gem	3.0
Melucta	ε Gem	3.0
Menchib	ξ Per	4.0
Menkalinan (Menkalina)	β Aur	1.9
Menkar (Mekab, Menkab)	α Cet	2.5
Menkar	λ Cet	4.7
Menkent	θ Cen	2.1
Menkib (Menkhib)	ξ Per	4.0
Menkib	β Peg	2.4
Merak	β UMa	2.4
Merez (Meres)	β Boo	3.6

Star name	Designation	Visual magnitude
Merga	38 Boo	5.8
Meridiana	α CrA	4.1
Merope	23 Tau	4.2
Mesarthim (Mesartim)	γ Ari	4.8
Metallah	α Tri	3.6
Miaplacidus (Maiaplacidus)	β Car	1.7
Mimosa	β Cru	1.3
Minelauva	β Vir	3.6
Minelauva	δ Vir	3.4
Minkar	ε Crv	3.0
Mintaka (Mintika)	δ Ori	2.2
Mira	o Cet	3.0
Mirach (Mirac, Merach)	β And	2.1
Mirach (Mirac, Mirak)	ε Boo	2.4
Miram	η Per	3.9
Mirfak (Mirphak, Mirzak)	α Per	1.8
Mirfak	κ Her	5.0
Mirza	γ CMa	4.1
Mirzam (Mirzim, Mirza)	β CMa	2.0
Misam	κ Per	4.0
Misam	λ Her	4.4
Mismar	α UMi	2.0
Mizar (Mirza, Mizat)	ζ UMa	2.4
Mizar	β And	2.1
Mizar	ε Boo	2.4
Monkar	α Cet	2.5
Mothallah	α Tri	3.4
Mufrid (Mufride, Muphrid, Muphride)	η Boo	2.7
Muliphain (Muliphein, Muliphen)	γ CMa	4.0
Murzim	β CMa	2.0
Muscida (Museida)	π² UMa	4.8
Muscida	o UMa	3.4
Nair al Saif	ι Ori	2.8
Nair al Zaurak	α Phe	2.4
Naos	ζ Pup	2.3
Nash (Nasl)	γ Sgr	3.1
Nashira	γ Cap	3.8
Nath	β Tau	1.6
Navi	ε Cas	3.4
Navigatoria	α UMi	2.0
Nekkar	β Boo	3.6
Nicolaus	α Del	3.8
Nihal (Nibal)	β Lep	2.8

Star name	Designation	Visual magnitude
Nodus I	ζ Dra	3.2
Nodus II (Nodus Secundus)	δ Dra	3.2
Nunki	σ Sgr	2.0
Nusakan	β CrB	3.1
Nushaba	γ Sgr	3.1
Oculus Boreus	ε Tau	3.5
Okda	α Psc	3.8
Osiris	α CMa	−1.5
Os Pegasi	ε Peg	2.4
Os Piscis Meridiani (Os Piscis Notii)	α PsA	1.2
Palilicium (Parilicium)	α Tau	0.9
Peacock	α Pav	1.9
Phad (Phacd)	γ UMa	2.5
Phakt (Phact, Phad, Phaet)	α Col	2.6
Phecda (Phegda, Phekda, Phekha)	γ UMa	2.4
Pherkad	γ UMi	3.1
Pherkad Major	γ UMi	3.1
Pherkad Minor	λ UMi	5.0
Pherkard	δ UMi	4.4
Phoenice	α UMi	2.0
Phurud	ζ CMa	3.0
Pishpai	μ Gem	2.9
Plaskett's Star	V640 Mon	6.1
Pleione	28 Tau	5.1
Polaris	α UMi	2.0
Polaris Australis	σ Oct	5.5
Pollux	β Gem	1.1
Porrima	γ Vir	2.8
Praecipua	46 LMi	3.8
Prima Giedi	α Cap	3.4
Procyon	α CMi	0.4
Propus	η Gem	3.3
Propus	ι Gem	3.8
Protrygetor	ε Vir	2.8
Proxima Centauri	α Cen C	11.2
Pulcherrima	ε Boo	2.4
Rana	δ Eri	3.7
Rana Secunda	β Cet	2.0
Rasalas (Ras al Asad)	μ Leo	4.1
Ras Algethi (Rasalgethi, Rasalegti)	α Her	3.5
Rasalhague (Ras Alhagua, Ras-al-hague)	α Oph	2.1
Ras al Mothallath (Ras al Muthallath)	α Tri	3.4
Ras Elased Australis	ε Leo	3.0

Star name	Designation	Visual magnitude
Ras Elased Boraelis	μ Leo	4.1
Ras Hammel	α Ari	2.0
Rastaban (Rastaben, Rasaben)	β Dra	2.4
Rastaban	γ Dra	2.2
Reda	γ Aql	2.8
Regor	γ Vel	1.8
Regulus	α Leo	1.4
Rescha	α Psc	3.8
Rex	α Leo	1.4
Rigel	β Ori	0.1
Rigil Kentaurus (Rigel Kentaurus, Rigil Kent, Rigel Kent)	α Cen	−0.3
Rigl al Awwa	μ Vir	3.9
Rotanev (Rotanen)	β Del	3.7
Ruchba (Rukbat al-dejajah)	o² Cyg	5.4
Ruchbah (Rucha)	δ Cas	2.7
Ruchbah	ε Cas	3.4
Rukbat (Rukbat al Rami)	α Sgr	4.1
Rutilicus	β Her	2.8
Sabik	η Oph	2.4
Saclateni (Sadatoni)	ζ Aur	3.8
Sadachbia (Sadalachbia)	γ Aqr	3.8
Sadalbari	μ Peg	3.5
Sadalmelek (Saad el Melik, Sadalmelik, Sadal Melik, Sadlamulk)	α Aqr	3.0
Sadalsuud (Saad el Sund, Sadalsud, Sadalsund, Sad es Saud)	β Aqr	2.9
Sadira	σ Sgr	2.0
Sadr (Sadir, Sador, Sadr el dedschadsche)	γ Cyg	2.2
Saidak	80 UMa	4.0
Saiph	κ Ori	2.1
Saiph	η Ori	3.4
Saiph	ι Ori	2.8
Salm	τ Peg	4.6
Sargas	θ Sco	1.9
Sarin	δ Her	3.1
Sartan (Sertan)	α Cnc	4.3
Scalovin	α Del	3.8
Sceptrum	53 Eri	4.0
Scheat (Sheat)	β Peg	2.4
Scheat	δ Aqr	3.3
Schedar (Schedir, Shedir)	α Cas	2.2
Scheddi (Sheddi)	δ Cap	3.0

Star name	Designation	Visual magnitude
Schemali (Shemali)	ι Cet	3.6
Scutulum	ι Car	2.2
Seat Alpheras	β Peg	2.4
Secunda Giedi	α² Cap	3.6
Segin	ε Cas	3.4
Seginus	γ Boo	3.0
Sham	α Sge	4.4
Shaula	λ Sco	1.6
Sheliak (Shelyak, Shiliak)	β Lyr	3.4
Sheratan (Sharatan)	β Ari	2.6
Singer	α Aur	0.1
Sirius	α CMa	−1.5
Sirrah (Sirah)	α And	2.1
Situla	κ Aqr	5.3
Skat	δ Aqr	3.3
Spica (Spica Virginis)	α Vir	1.0
Sterope	21 Tau	5.8
Sualocin (Svalocin)	α Del	3.9
Subra	o Leo	3.8
Suha	80 UMa	4.0
Suhail (Suhel)	α Car	−0.7
Suhail (Suhail al Wazn)	λ Vel	2.2
Suhail al Mulif	γ Vel	1.8
Suhail Hadar	ζ Pup	2.3
Sulaphat (Sulafat)	γ Lyr	3.3
Superba	Y CVn	5.0
Syrma	ι Vir	4.5
Tabit	π³ Ori	3.3
Tabit (Thabit)	υ Ori	4.6
Talitha (Talita)	ι UMa	3.1
Tania Australis	μ UMa	3.1
Tania Boraelis	λ UMa	3.5
Tarazed (Tarazad)	γ Aql	2.7
Tarf	β Cnc	3.5
Taygeta (Tayeta, Taygete)	19 Tau	4.4
Tegmen (Tegmine)	ζ Cnc	4.7
Tejat	μ Gem	3.2
Tejat Posterior	μ Gem	2.9
Tejat Prior	η Gem	3.4
Terebellum	59 Sgr	4.5
Theemin (Theemim)	υ² Eri	3.8
Thuban	α Dra	3.7
Toliman (Tolimann)	α Cen	−0.3
Torcularis Septentrionalis	o Psc	4.3

Star name	Designation	Visual magnitude
Tramontana	α UMi	2.0
Tsih	γ Cas	2.5
Tureis (Turais)	ι Car	2.3
Tyl	ε Dra	4.0
Unuk al Hai (Unuk, Unukalhai, Unuk Elhaia)	α Ser	2.7
Urkab Posterior	β² Sgr	4.3
Urkab Prior	β¹ Sgr	4.0
Variabilis Coronae	R CrB	5.9
Vega	α Lyr	0.0
Venabulum	μ Boo	4.3
Venator	β Del	3.6
Vespertilio	α Sco	1.2
Vildiur	δ UMi	4.4
Vindemiatrix (Vendemiatrix, Vindemiator)	ε Vir	2.8
Wasat	δ Gem	3.5
Wasn (Wazn)	β Col	3.1
Wezen (Wesen)	δ CMa	1.8
Yed (Yad)	δ Oph	3.0
Yed Posterior	ε Oph	3.3
Yed Prior	δ Oph	2.7
Yildun (Yilduz)	δ UMi	4.4
Yilduz	α UMi	2.0
Zania (Zaniah)	η Vir	4.1
Zarijan	β Vir	3.8
Zaurak (Zaurac, Zaurack)	γ Eri	3.0
Zavijava (Zavijah, Zavyava, Zawijah)	β Vir	3.8
Zenith Star	γ Dra	2.2
Zibel (Zibal)	ζ Eri	4.9
Zosma (Zosca, Zozca, Zozma)	δ Leo	2.6
Zubenelakrab (Zuben Elakrab, Zuben el Hakrabi)	γ Lib	4.0
Zubenelakribi (Zuben Elakribi)	δ Lib	4.9v[b]
Zuben el Chamali (Zubenelg)	β Lib	2.7
Zubenelgenubi (Zuben Elgenubi, Zuben el Genubi)	α Lib	2.9
Zuben Elgenubi (Zuben el Genubi)	γ Sco	3.3
Zubeneschamali (Zuben Elschemali, Zubenesch)	β Lib	2.6
Zubenhakrabi (Zuben Hakrabi, Zuben Hakraki)	γ Lib	5.3
Zuben Hakrabi	γ Sco	3.3
Zuben Hakrabi	η Lib	5.4

Star name	Designation	Visual magnitude
Zuben Hakrabi	ν Lib	5.2
Zubra	δ Leo	2.6
Zujj al Nushshabah	γ Sgr	3.0

[a] Combination of the following two entries
[b] v indicates a variable star

Sun signs – the constellations of the zodiac

1 TRADITIONAL DATES

Aries	21 Mar – 19 Apr	(30 days)
Taurus	20 Apr – 20 May	(31 days)
Gemini	21 May – 21 Jun	(32 days)
Cancer	22 Jun – 22 Jul	(31 days)
Leo	23 Jul – 22 Aug	(31 days)
Virgo	23 Aug – 22 Sep	(31 days)
Libra	23 Sep – 23 Oct	(31 days)
Scorpius	24 Oct – 21 Nov	(29 days)
Sagittarius	22 Nov – 21 Dec	(30 days)
Capricornus	22 Dec – 19 Jan	(29 days)
Aquarius	20 Jan – 18 Feb	(30 days)
Pisces	19 Feb – 20 Mar	(30/31 days)

2 ACTUAL DATES

Pisces	12 Mar – 18 Apr	(38 days)
Aries	19 Apr – 13 May	(25 days)
Taurus	14 May – 19 Jun	(37 days)
Gemini	20 Jun – 20 Jul	(31 days)
Cancer	21 Jul – 9 Aug	(20 days)
Leo	10 Aug – 15 Sep	(37 days)
Virgo	16 Sep – 30 Oct	(45 days)
Libra	31 Oct – 22 Nov	(23 days)
Scorpius	23 Nov – 29 Nov	(7 days)
Ophiuchus	30 Nov – 17 Dec	(18 days)
Sagittarius	18 Dec – 18 Jan	(32 days)
Capricornus	19 Jan – 15 Feb	(28 days)
Aquarius	16 Feb – 11 Mar	(24/25 days)

Note: 45 dates are in agreement with the traditional dates.

The visibility of the constellations

Constellation	Whole constellation visible from these Earth latitudes	Whole constellation invisible from these Earth latitudes
Andromeda	N of −37°	S of −69°
Antlia	S of +50°	N of +66°
Apus	S of +7°	N of +23°
Aquarius	between +65° and −87°	portions visible worldwide
Aquila	between +78° and −71°	portions visible worldwide
Ara	S of +22°	N of +45°
Aries	N of −59°	S of −80°
Auriga	N of − 34°	S of −62°
Boötes	N of −35°	S of −83°
Caelum	S of +41°	N of +63°
Camelopardalis	N of −5°	S of −37°
Cancer	N of −57°	S of −83°
Canes Venatici	N of −37°	S of −62°
Canis Major	S of +57°	N of +79°
Canis Minor	N of −77°	portions visible worldwide
Capricornus	S of +62°	N of +82°
Carina	S of +15°	N of +39°
Cassiopeia	N of −12°	S of −44°
Centaurus	S of +25°	N of +60°
Cepheus	N of −1°	S of −39°
Cetus	S of +65°	portions visible worldwide
Chamaeleon	S of +7°	N of +15°
Circinus	S of +20°	N of +36°
Columba	S of +47°	N of +63°
Coma Berenices	N of −56°	S of −77°
Corona Australis	S of +44°	N of +53°
Corona Borealis	N of −50°	S of −64°
Corvus	S of +65°	N of +79°
Crater	S of +65°	N of +84°
Crux	S of +25°	N of +35°
Cygnus	N of −29°	S of −62°
Delphinus	N of −69°	S of −88°
Dorado	S of +20°	N of +41°
Draco	N of −4°	S of −42°
Equuleus	N of −77°	S of −88°
Eridanus	S of +32°	portions visible worldwide
Fornax	S of +50°	N of +64°
Gemini	N of −55°	S of −80°

Constellation	Whole constellation visible from these Earth latitudes	Whole constellation invisible from these Earth latitudes
Grus	S of +33°	N of +53°
Hercules	N of −39°	S of −86°
Horologium	S of +23°	N of +50°
Hydra	between +55° and −83°	portions visible worldwide
Hydrus	S of +8°	N of +32°
Indus	S of +15°	N of +45°
Lacerta	N of −33°	S of −55°
Leo	between +84° and −57°	portions visible worldwide
Leo Minor	N of −48°	S of −67°
Lepus	S of +63°	N of +79°
Libra	S of +60°	portions visible worldwide
Lupus	S of +35°	N of +60°
Lynx	N of −28°	S of −57°
Lyra	N of −42°	S of −65°
Mensa	S of +5°	N of +20°
Microscopium	S of +45°	N of +62°
Monoceros	between +79° and −78°	portions visible worldwide
Musca	S of +15°	N of +26°
Norma	S of +30°	N of +48°
Octans	S of +00°	N of +15°
Ophiuchus	between +60° and −76°	portions visible worldwide
Orion	between +79° and −67°	portions visible worldwide
Pavo	S of +15°	N of +33°
Pegasus	N of −54°	S of −88°
Perseus	N of −31°	S of −59°
Phoenix	S of +32°	N of +50°
Pictor	S of +26°	N of +47°
Pisces	between +83° and −57°	portions visible worldwide
Piscis Austrinus	S of +53°	N of +65°
Puppis	S of +39°	N of +79°
Pyxis	S of +53°	N of +73°
Reticulum	S of +23°	N of +37°
Sagitta	N of −69°	S of −74°
Sagittarius	S of +45°	N of +78°
Scorpius	S of +44°	N of +82°
Sculptor	S of +50°	N of +65°
Scutum	S of +74°	N of +86°
Serpens	between +74° and −64°	portions visible worldwide
Sextans	between +79° and −83°	portions visible worldwide
Taurus	N of −59°	portions visible worldwide
Telescopium	S of +33°	N of +45°
Triangulum	N of −53°	S of −65°

Constellation	Whole constellation visible from these Earth latitudes	Whole constellation invisible from these Earth latitudes
Triangulum Australe	S of +20°	N of +30°
Tucana	S of +14°	N of +33°
Ursa Major	N of −17°	S of −61°
Ursa Minor	N of 00°	S of −25°
Vela	S of +33°	N of +53°
Virgo	between +68° and −76°	portions visible worldwide
Volans	S of +15°	N of +26°
Vulpecula	N of −61°	S of −71°

The number of visible stars in the constellations

Constellation	Number of stars brighter than 2.4	Number of stars with brightness between 2.4 and 4.4	Number of stars with brightness between 4.4 and 5.5	Total
Andromeda	3	14	37	54
Antlia	0	1	8	9
Apus	0	4	6	10
Aquarius	0	18	38	56
Aquila	1	12	34	47
Ara	0	8	11	19
Aries	1	4	23	28
Auriga	2	9	36	47
Boötes	2	12	39	53
Caelum	0	1	3	4
Camelopardalis	0	5	40	45
Cancer	0	4	19	23
Canes Venatici	0	2	13	15
Canis Major	5	13	38	56
Canis Minor	1	3	9	13
Capricornus	0	10	21	31
Carina	4	20	53	77
Cassiopeia	3	8	40	51
Centaurus	6	31	64	101
Cepheus	1	14	42	57
Cetus	1	14	43	58
Chamaeleon	0	5	8	13
Circinus	0	2	8	10
Columba	0	7	17	24
Coma Berenices	0	3	20	23
Corona Australis	0	3	18	21
Corona Borealis	1	4	17	22
Corvus	0	6	5	11
Crater	0	3	8	11
Crux	3	6	11	20
Cygnus	3	20	56	79
Delphinus	0	5	6	11
Dorado	0	4	11	15
Draco	1	16	62	79

Constellation	Number of stars brighter than 2.4	Number of stars with brightness between 2.4 and 4.4	Number of stars with brightness between 4.4 and 5.5	Total
Equuleus	0	1	4	5
Eridanus	1	29	49	79
Fornax	0	2	10	12
Gemini	3	16	28	47
Grus	2	8	14	24
Hercules	0	24	61	85
Horologium	0	1	9	10
Hydra	1	19	51	71
Hydrus	0	5	9	14
Indus	0	4	9	13
Lacerta	0	3	20	23
Leo	3	15	34	52
Leo Minor	0	2	13	15
Lepus	0	10	18	28
Libra	0	7	28	35
Lupus	1	20	29	50
Lynx	0	5	26	31
Lyra	1	8	17	26
Mensa	0	0	8	8
Microscopium	0	0	15	15
Monoceros	0	6	30	36
Musca	0	6	13	19
Norma	0	1	13	14
Octans	0	3	14	17
Ophiuchus	2	20	33	55
Orion	7	19	51	77
Pavo	1	10	17	28
Pegasus	1	15	41	57
Perseus	1	22	42	65
Phoenix	1	8	18	27
Pictor	0	2	13	15
Pisces	0	11	39	50
Piscis Austrinus	1	4	10	15
Puppis	1	19	73	93
Pyxis	0	3	9	12
Reticulum	0	3	8	11
Sagitta	0	4	4	8
Sagittarius	2	18	45	65
Scorpius	6	19	37	62
Sculptor	0	3	12	15

Constellation	Number of stars brighter than 2.4	Number of stars with brightness between 2.4 and 4.4	Number of stars with brightness between 4.4 and 5.5	Total
Scutum	0	2	7	9
Serpens	0	13	23	36
Sextans	0	0	5	5
Taurus	2	26	70	98
Telescopium	0	2	15	17
Triangulum	0	3	9	12
Triangulum Australe	1	4	7	12
Tucana	0	4	11	15
Ursa Major	6	14	51	71
Ursa Minor	2	5	11	18
Vela	3	18	55	76
Virgo	1	15	42	58
Volans	0	6	8	14
Vulpecula	0	1	28	29
Totals	88	779	2180	3047

2 Constellations

Part 2 attempts to group much of the data about each constellation in an easily accessible form. All information, with the exceptions of non-traditional mythology and interesting facts, comes from the lists of Part 1.

The number in parenthesis after each of the bright and near stars show the star's rank as it is found on the lists of the 200 brightest stars and 200 nearest stars. For example η Boo is the 106th brightest and the 186th nearest star. A similar rank is given for the overall brightness of the constellation (Boötes is the 59th brightest constellation). The identification in parenthesis following the name of each star is that star's designation, usually the Greek letter assigned by Bayer. Only the most popular star name, with perhaps one or two dominant variants, was carried over from the Star names list in part 1. Thus, because Izar and Pulcherrima are both common forms of the star ε Boo, both are listed.

The non-traditional mythology section contains anecdotal references to the constellations. Some of these tales are quite new. As may be expected, there is a bias toward the bright constellations which can be seen by northern hemisphere observers.

Finally, details for each constellation end with a section of interesting facts. At least one fact has been listed for each constellation, although in certain cases finding even one fact was a difficult chore.

A note about the antique star maps. The constellations which cover the sky have a rich and beautiful history. Nothing illustrates this better than the numerous and diverse star maps which have been created through the centuries. On the page opposite each constellation's details, an antique star map will be found, selected to show some of the variety which has been brought into existence. All maps which depict 'invented' constellations (those which have been created since the time of Ptolemy) show either the first appearance of the constellation on any star map or the oldest surviving illustration.

A note about the star charts. On the page opposite each constellation's details may be found a star chart featuring that constellation and all bordering constellations. Individual stars are shown down to visual magnitude 5.6. Most constellations are displayed using Mercator projection. Those near the poles, however, are represented by a spherical projection, as this produces a map which better represents what is seen in the sky. The lines separating the constellations are the official IAU boundaries, first established in 1928.

Andromeda

Meaning:	The Princess of Ethiopia
Pronunciation:	an draw' meh duh
Abbreviation:	And
Possessive form:	Andromedae (an drom' uh die)
Asterisms:	The Baseball Diamond, Frederik's Glory, The Great Square, The Large Dipper The Three Guides

Bordering constellations: Cassiopeia, Lacerta, Pegasus, Perseus, Pisces, Triangulum

Overall brightness:	7.476 (37)
Central point:	RA = 0h46m Dec. = +37°

Directional extremes: N = +53°   S = +21°   E = 2h36m   W = 22h56m

Messier objects:	M31, M32, M110
Meteor showers:	annual Andromedids (3 Oct) Andromedids (27 Nov)

Midnight culmination date: 9 Oct

Bright stars:	α (51), β (52), γ^l (69)
Named stars:	Adhil (ξ), Alamak (γ^l), Almach (γ), Alpheratz (α), Mirach (β), Sirrah (α)
Near stars:	Ross 248 (9), Groombridge 34A-B (17)
Size:	722.28 square degrees (1.751% of the sky)
Rank in size:	19

Solar conjunction date: 3 Apr

Visibility:	completely visible from latitudes: N of –37° completely invisible from latitudes: S of –69°
Visible stars:	(number of stars brighter than magnitude 5.5): 54

Non-traditional 'mythology': β And and γ And, along with the four stars in the 'great square' of Pegasus (α And, α Peg, β Peg, γ Peg) and α Per, comprise what some have termed the 'Giant Dipper.' This 'dipper,' like its 'big' northern counterpart, also has a pair of stars at the end of the bowl which point to the north celestial pole. If β Per (presumably at maximum) is used instead of α, one can even note a "bend" in the handle!

Interesting facts: (1) Alpheratz (α And) has only recently (1928) been assigned to Andromeda. Since ancient times it has been a common star with Pegasus lying, as it does, at the northeast corner of the Great Square. It was also known as δ Peg, but did not have any other common name which specifically linked it to Pegasus.

(2) The Andromeda Galaxy, M31, is often referred to as the most distant visible naked-eye object. Some observers, however, are able to detect M33, the spiral galaxy in Triangulum which is about one and one-third magnitudes fainter than M31.

(3) M31 is 2.3 million light years away and is approaching us at approximately 300 km/sec. It is the closest known spiral galaxy.

(4) Almach (γ And) is a multiple star system. Three stars may be telescopically detected, although small instruments will only show the two brightest. These two show a remarkable contrast in color, as one is orange and the other blue.

Andromeda
Coronelli, Vincenzo.
Epitome
Cosmografica...,
Cologne, 1693.

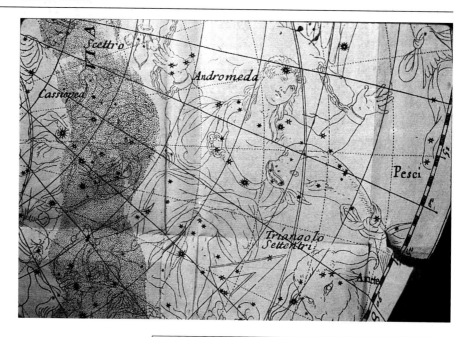

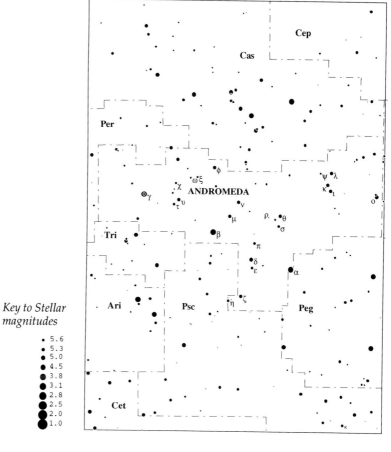

*Key to Stellar
magnitudes*

· 5.6
· 5.3
· 5.0
● 4.5
● 3.8
● 3.1
● 2.8
● 2.5
● 2.0
● 1.0

Antlia

Meaning:	The Air Pump
Pronunciation:	ant' lee ah
Abbreviation:	Ant
Possessive form:	Antliae (ant' lee eye)
Asterisms:	none
Bordering constellations:	Centaurus, Hydra, Pyxis, Vela
Overall brightness:	3.767 (83)
Central point:	RA = 10h14m Dec. = −32°
Directional extremes:	N = −24° S = −40° E = 11h03m W = 9h25m
Messier objects:	none
Meteor showers:	none
Midnight culmination date:	24 Feb
Bright stars:	none
Named stars:	none
Near stars:	none
Size:	238.90 square degrees (0.579% of the sky)
Rank in size:	62
Solar conjunction date:	26 Aug
Visibility:	completely visible from latitude: S of +50°
	completely invisible from latitudes: N of +66°
Visible stars:	(number of stars brighter than magnitude 5.5): 9
Interesting facts:	(1) This was one of the 14 constellations invented by Lacaille during his stay at the Cape of Good Hope in 1751–52. Lacaille created this constellation to honor the invention of the air pump by Robert Boyle.

Antlia (labeled 'la Machine Pneumatique' on this map) Lacaille, Nicolas Louis de. Planisphere contenant les Constellations Celestes, found in Mémoires Académie Royale des Sciences, *Paris, 1752 (published in 1756). This constellation was invented by Lacaille and the photo shows its first appearance on any star map.*

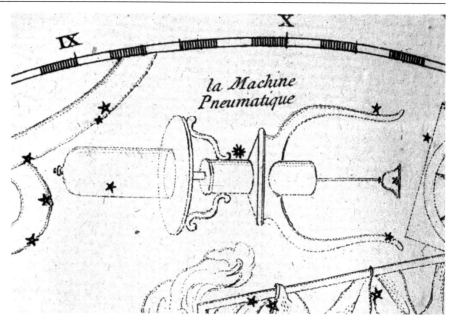

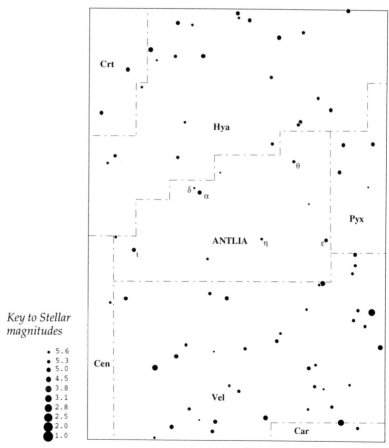

Key to Stellar magnitudes

- 5.6
- 5.3
- 5.0
- 4.5
- 3.8
- 3.1
- 2.8
- 2.5
- 2.0
- 1.0

Apus

Meaning:	The Bird of Paradise
Pronunciation:	ape' us
Abbreviation:	Aps
Possessive form:	Apodis (ap' oh diss)
Asterisms:	none
Bordering constellations:	Ara, Chamaeleon, Circinus, Musca, Octans, Pavo, Triangulum Australe
Overall brightness:	4.847 (76)
Central point:	R.A. = 16h01m Dec. = −75°
Directional extremes:	N = −67° S = +21° E = 18h17m W = 13h45m
Messier objects:	none
Meteor showers:	none
Midnight culmination date:	21 May
Bright stars:	none
Named stars:	none
Near stars:	LFT 1338 (200)
Size:	206.32 square degrees (0.500% of the sky)
Rank in size:	67
Solar conjunction date:	22 Nov
Visibility:	completely visible from latitudes: S of N +7°
	completely invisible from latitudes: N of +23°
Visible stars:	(number of stars brighter than magnitude 5.5): 10
Interesting facts:	(1) This is one of 11 constellations invented by Pieter Dirksz Keyser and Frederick de Houtman, during the years 1595–7.

Apus (labeled 'Apis Indica' on this map) Bayer, Johann. Uranometria, *Augsburg, 1603. This constellation was invented by de Houtman and Keyser in 1596. It was first illustrated on a globe by Plancius, which has not survived. This photo from Bayer's map, therefore, shows the earliest existing picture of this constellation.*

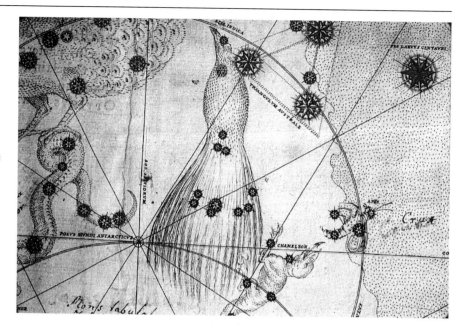

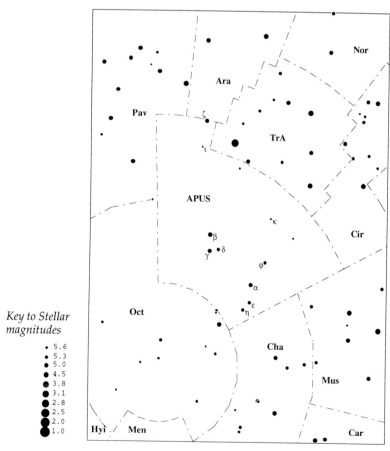

Key to Stellar magnitudes

- 5.6
- 5.3
- 5.0
- 4.5
- 3.8
- 3.1
- 2.8
- 2.5
- 2.0
- 1.0

Aquarius

Meaning:	The Water Bearer
Pronunciation:	uh qwayr' ee us
Abbreviation:	Aqr
Possessive form:	Aquarii (ah kwayr' ee ee)
Asterisms:	The Water Jar
Bordering constellations:	Aquila, Capricornus, Cetus, Delphinus, Equuleus, Pegasus, Pisces, Piscis Austrinus, Sculptor
Overall brightness:	5.715 (65)
Central point:	RA = 22h15m Dec. = –11°
Directional extremes:	N = +3° S = –25° E = 23h54m W = 20h36m
Messier objects:	M2, M72, M73
Meteor showers:	η Aquarids (3 May)
	S. δ Aquarids (29 Jul)
	S. ι Aquarids (5 Aug)
	N. δ Aquarids (12 Aug)
	N. ι Aquarids (20 Sep)
	κ Aquarids (21 Sep)
Midnight culmination date:	25 Aug
Bright stars:	β (152), α (160)
Named stars:	Albali (ε), Ancha (θ), Sadachbia (γ), Sadalmelek (α), Sadalsuud (β), Situla (κ), Skat (δ)
Near stars:	L 789–6 (11), Ross 780 (35), LFT 1754 (105), Wolf 1329 (145), BD–21°6267A–B (161), BD+0°4810 (168), LFT 1699–1700 (178)
Size:	979.85 square degrees (2.375% of the sky)
Rank in size:	10
Solar conjunction date:	24 Feb
Visibility:	completely visible from latitudes: +65° to – 87°
	portions visible worldwide
Visible stars:	(number of stars brighter than magnitude 5.5): 56

Non-traditional 'mythology': It is possible to make a rough outline map of South America using the stars θ, λ, τ, δ, 88, and ι in this constellation.

Interesting facts:
(1) On 23 September 1846, the planet Neptune was discovered within the borders of this constellation, about 1° north of ι Aqr, by the German astronomer Galle at the Berlin Observatory. Its position had been predicted by a French astronomer, Urbain Leverrier. Later, the English astronomer, John Couch Adams was also given credit for predicting the location of Neptune.

(2) The oft-mentioned 'Age of Aquarius' will occur when the vernal equinox moves from its present position in Pisces into Aquarius. This movement is caused by the Earth's precessional motion. Don't look for worldwide peace and understanding too soon, however – this momentous event is still about 800 years away!

(3) Midway between the stars ι Aqr and α PsA (Fomalhaut), we find the wonderful planetary nebula known as the Helix Nebula. At a distance of approximately 700 light years, it is the nearest of all planetary nebulae. As it is so near, it also seems to be the largest, covering $\frac{1}{4}°$ in area. Experience has demonstrated that the best views of this object of low surface brightness are through binoculars which give ×10–×20 magnification.

Aquarius
Flamsteed, John. Atlas
Coelestis, *London,*
1729.

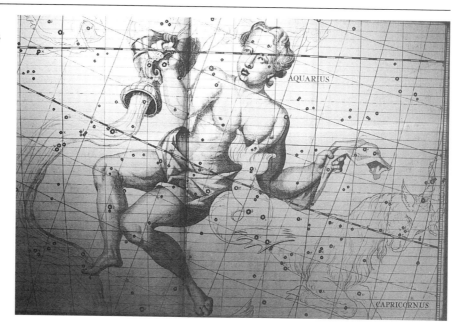

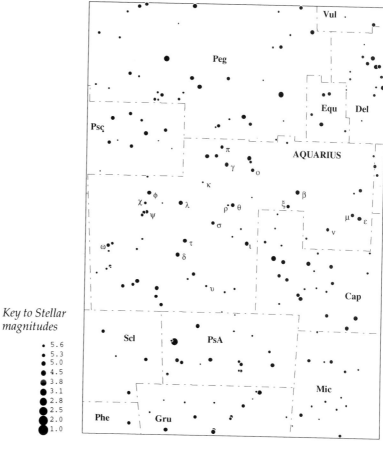

Key to Stellar
magnitudes

- 5.6
- 5.3
- 5.0
- 4.5
- 3.8
- 3.1
- 2.8
- 2.5
- 2.0
- 1.0

Aquila

Meaning:	The Eagle
Pronunciation:	ak' will uh
Abbreviation:	Aql
Possessive form:	Aquilae (ak' will eye)
Asterisms:	The Family, The Summer Triangle
Bordering constellations:	Aquarius, Capricornus, Delphinus, Hercules, Ophiuchus, Sagitta, Sagittarius, Scutum, Serpens
Overall brightness:	7.203 (41)
Central point:	RA = 19h37m Dec. = +3.5°
Directional extremes:	N = +19° S = −12° E = 20h36m W = 18h38m
Messier objects:	none
Meteor showers:	none
Midnight culmination date:	16 Jul
Bright stars:	α (12), γ (114), ζ (165)
Named stars:	Alshain (β), Altair (α), Deneb Okab (δ), Reda (γ), Tarazed (γ)
Near stars:	α (40), Wolf 1055 A (56), LTT 7658–7659 (177)
Size:	652.47 square degrees (1.582% of the sky)
Rank in size:	22
Solar conjunction date:	15 Jan
Visibility:	completely visible from latitudes: +78° to −71° portions visible worldwide
Visible stars:	(number of stars brighter than magnitude 5.5): 47

Interesting facts:

(1) Nova Aquilae, one of the most famous novae in recent times, blazed forth on the night of 8 June 1918, about 7° northwest of λ Aql. It was the brightest nova to appear since Kepler's Nova in 1604.

(2) Another nova, which apparently equalled Venus in brightness, appeared near α Aql in AD 389. It was visible for three weeks.

(3) Altair (α Aql) is notable for its extremely rapid rotation. From studies of its spectrum, it has been shown that Altair spins once every 6 1/2 hours. Such a rapid motion must distort this star's shape tremendously. It has been estimated that the equatorial diameter of Altair is twice its polar diameter.

(4) The star with the lowest measured luminosity is found within Aquila. Just to the northwest of the star δ Aql is a 9th magnitude star whose companion is known as 'Van Biesbroeck's Star.' This star has an absolute magnitude of +19.3. If Van Biesbroeck's Star were placed side-by-side with our Sun (an average star whose absolute magnitude is +4.6), it's brightness would only be 1/758 000 of the Sun's.

Aquila
Bode, Johann Elert.
Uranographia Sive
Astrorum
Descriptio, *Berlin,*
1801.

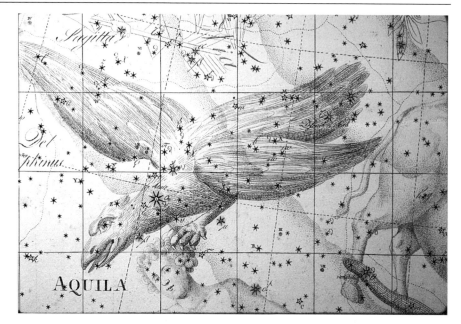

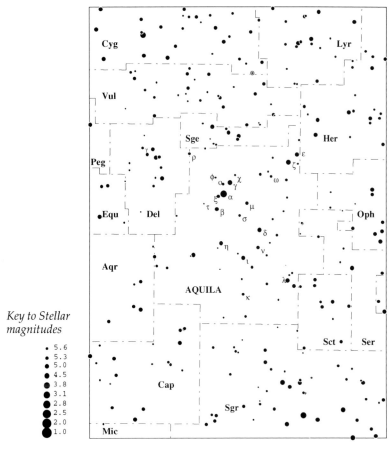

Key to Stellar
magnitudes

145

Ara

Meaning:	The Altar
Pronunciation:	air′ uh
Abbreviation:	Ara
Possessive form:	Arae (air′ eye)
Asterisms:	none

Bordering constellations: Apus, Corona Australis, Norma, Pavo, Scorpius, Telescopium, Triangulum Australe

Overall brightness:	8.015 (34)
Central point:	RA = 17h18m Dec. = −56.5°

Directional extremes: N = −45°   S = −68°   E = 18h06m   W = 16h31m

Messier objects:	none
Meteor showers:	none

Midnight culmination date: 10 Jun

Bright stars:	β (137), α (158), ζ (195)
Named stars:	none
Near stars:	LFT 1351 (30), 41 Ara A-B (103), LFT 1297 (167)
Size:	237.06 square degrees (0.575% of the sky)
Rank in size:	63

Solar conjunction date: 9 Dec

Visibility:	completely visible from latitudes: S of +22° completely invisible from latitudes: N of +45°
Visible stars:	(number of stars brighter than magnitude 5.5): 19

Interesting facts: (1) NGC 6397, with a visual magnitude of 7.5, is probably the nearest globular cluster to our solar system. It lies at a distance of only 8400 light years.

Ara
Rost, Johann
Leonhard. Atlas
Portatilis Coelestis,
Nuremburg, 1723.

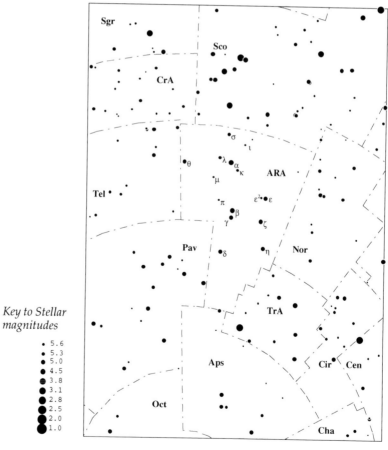

Key to Stellar
magnitudes

Aries

Meaning:	The Ram
Pronunciation:	air' eeze
Abbreviation:	Ari
Possessive form:	Arietis (air ee' ay tiss)
Asterisms:	The Northern Fly

Bordering constellations: Cetus, Perseus, Pisces, Taurus, Triangulum

Overall brightness: 6.344 (53)

Central point: RA = 2h35m   Dec. = +20.5°

Directional extremes: N = +31°   S = +10°   E = 3h27m   W = 1h44m

Messier objects: none

Meteor showers: Daytime Arietids (7 Jun)

δ Arietids (11 Dec)

Midnight culmination date: 30 Oct

Bright stars: α (47), β (101)

Named stars: Botein (δ), Hamal (α), Mesarthim (γ), Sheratan (β)

Near stars: LFT 171 (34), LFT 215 (97), Ross 556 (114)

Size: 441.39 square degrees (1.07% of the sky)

Rank in size: 39

Solar conjunction date: 31 Apr

Visibility: completely visible from latitudes: N of –59°

completely inivisible from latitudes: S of –80°

Visible stars: (number of stars brighter than magnitude 5.5):   28

Interesting facts: (1) γ Arietis was one of the first double stars to be detected. It was discovered accidentally by Robert Hooke in 1664. He had been telescopically following a comet at the time.

(2) About the year 27 BC, the 'First Point of Aries' moved from the constellation Aries into Pisces. This point, also known as the vernal equinox, marks the position of the sun on the ecliptic where it crosses the celestial equator heading north.

Aries
Seller, John. Atlas
Coelestis..., *London,*
1700.

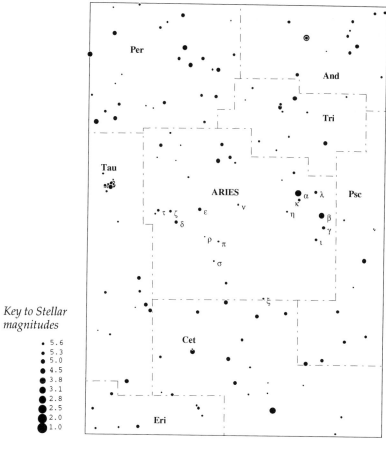

Key to Stellar
magnitudes

· 5.6
· 5.3
· 5.0
● 4.5
● 3.8
● 3.1
● 2.8
● 2.5
● 2.0
● 1.0

149

Auriga

Meaning:	The Charioteer
Pronunciation:	or eye' guh
Abbreviation:	Aur
Possessive form:	Aurigae (or eye' guy)
Asterisms:	The Heavenly G, The Kids, The Winter Octagon, The Winter Oval

Bordering constellations: Camelopardalis, Gemini, Lynx, Perseus, Taurus

Overall brightness: 7.149 (43)

Central point: RA = 6h01m   Dec. = +42°

Directional extremes: N = +56°   S = +28°   E = 7h27m   W = 4h35m

Messier objects: M36, M37, M38

Meteor showers: Aurigids (1 Sep)

Midnight culmination date: 21 Dec

Bright stars: α (6), β (40), ι (108), ε (163), η (199)

Named stars: Alhajoth (α), Almaaz (ε), Capella (α), Hasseleh (ι), Hoedus I (ζ), Hoedus II (η), Maaz (ε)

Near stars: Ross 986 (60), BD+53°935 (129)

Size: 657.44 square degrees (1.594% of the sky)

Rank in size: 21

Solar conjunction date: 22 Jun

Visibility: completely visible from latitudes: N of –34°
completely invisible from latitudes: S of –62°

Visible stars: (number of stars brighter than magnitude 5.5):   47

Interesting facts: (1) β Tau was originally a 'common' star with this constellation. On some star maps it was also called γ Aur. When the IAU officially designated constellation boundaries in 1928, the star was permanently assigned to Taurus. Apparently this was done because the star serves a more prominent role in Taurus, being a tip of one of the Bull's horns. It has been rumored, however, that the star was assigned to Taurus simply because β comes before γ in the Greek alphabet.

(2) ε Aur is a very unusual type of star. Known as an eclipsing binary, two stars are represented in this system. Every 27 years, the dark companion blots out much of the light from the primary in an eclipse that lasts a full year. The magnitude of the primary falls from 3.0 to 3.8, a twofold decrease of brightness. Of the many explanations that have been offered for this system, the most probable suggests that the dark companion is actually a cloud of pre-stellar matter, co-orbiting with ε Aur.

Auriga
Sherburne, Edward.
The Sphere of
Marcus Manilius
made an English
poem: with
annotations and an
astronomical
appendix, *London,*
1675.

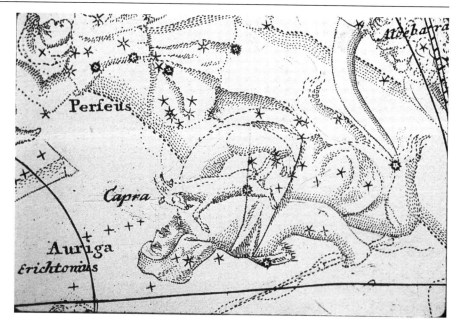

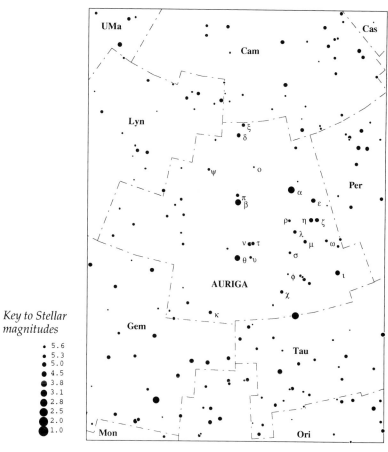

Key to Stellar
magnitudes

151

Boötes

Meaning:	The Bear Driver
Pronunciation:	bow owe' teez
Abbreviation:	Boo
Possessive form:	Boötis (bow owe' tiss)
Asterisms:	The Diamond [of Virgo], The Ice Cream Cone, The Kite, The Spring Triangle, The Trapezoid
Bordering constellations:	Canes Venatici, Coma Berenices, Corona Borealis, Draco, Hercules, Serpens, Ursa Major, Virgo
Overall brightness:	5.845 (59)
Central point:	RA = 14h40m Dec. = +31°
Directional extremes:	N = +55° S = +7° E = 15h47m W = 13h33m
Messier objects:	none
Meteor showers:	Quadrantids (3 Jan) φ Boötids (1 May)
	α Boötids (28 Apr) June Bootids (28 Jun)
Midnight culmination date:	2 May
Bright stars:	α (4), ε (78), η (106), γ (175)
Named stars:	Arcturus (α), Haris (γ), Izar (ε), Merez (β), Merga (38), Mufrid (η), Nekkar (β), Pulcherrima (ε), Seginus (γ)
Near stars:	Wolf 498 (39), ξ Boo A-B (81), η Boo (186)
Size:	906.83 square degrees (2.198% of the sky)
Rank in size:	13
Solar conjunction date:	1 Nov
Visibility:	completely visible from latitudes: N of –35°
	completely invisible from latitudes: S of –83°
Visible stars:	(number of stars brighter than magnitude 5.5): 53

Non-traditional 'mythology': Although many have seen a kite-shaped figure within the main body of this constellation, I have heard it described somewhat differently. α Boo, it is said, lies at the bottom of an ice-cream cone. The remainder of the (sugar) cone is formed by the stars ε, δ, γ, and ρ of this constellation. β Boo marks the top of the scoop of ice cream, which rests comfortably upon the cone. It is further stated that this particular object was at one time a two-scoop cone; however, as Boötes is near its highest point in the early evening during the hottest days of summer, the second scoop has melted, slipped off, and is now found just to the east of the cone as the constellation Corona Borealis.

Interesting facts: (1) Arcturus (α Boo) was the first star to be observed in the daytime, in 1635. It was also seen 'through' (or behind) the heads of two bright comets: the comet of 1618 and Donati's comet of 1858. In fact, on 28 September 1858, the first photograph of a comet (Donati's) was taken near Arcturus.

(2) The heat of Arcturus has been carefully measured and has been found to be equal to that of a standard candle at the distance of 5 miles (8 kilometres).

(3) The light from Arcturus opened the 1933 World Exposition in Chicago. It was then believed that the distance to the star was 40 light years. 40 years before, there had been another great Exposition in Chicago. It was thought appropriate that the light which left Arcturus in 1893, during the last great fair, would open the 1933 event. The light from this star was collected by telescope and focused on a photocell which turned on the lights during the first night of the Exposition. The distance to Arcturus has since been correctly revised to approximately 36 light years.

(4) ε Boo, often called Izar has another common name, bestowed upon it by F. G. W. Struve. He called it Pulcherrima, which is Latin for 'most beautiful.' This refers to its telescopic appearance as a colorful double star, the components being orange and blue.

Boötes
Cellarius, Andreas.
Harmonia
Macrocosmica sev
Atlas Universalis et
Novus, *Amsterdam,*
1661.

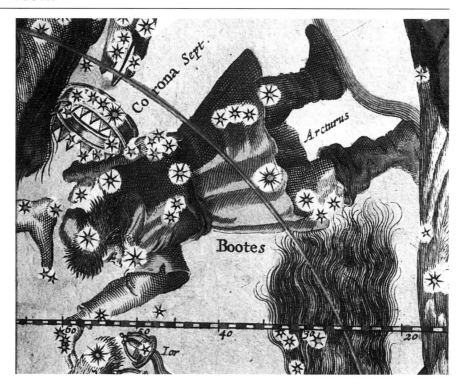

Key to Stellar
magnitudes

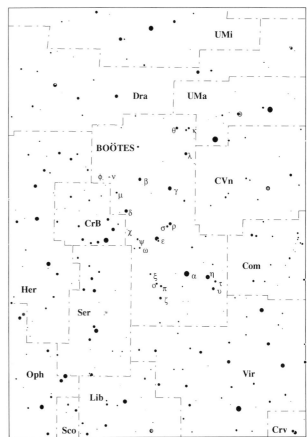

153

Caelum

Meaning:	The Sculptor's Chisel
Pronunciation:	see' lum
Abbreviation:	Cae
Possessive form:	Caeli (see' lee)
Asterisms:	none
Bordering constellations:	Columba, Dorado, Eridanus, Horologium, Lepus, Pictor
Overall brightness:	3.204 (85)
Central point:	RA = 4h40m Dec. = −38°
Directional extremes:	N = −27° S = −49° E = 5h03m W = 4h18m
Messier objects:	none
Meteor showers:	none
Midnight culmination date:	1 Dec
Bright stars:	none
Named stars:	none
Near stars:	none
Size:	124.86 square degrees (0.303% of the sky)
Rank in size:	81
Solar conjunction date:	2 Jun
Visibility:	completely visible from latitudes: S of +41° completely invisible from latitudes: N of +63°
Visible stars:	(number of stars brighter than magnitude 5.5): 4
Interesting facts:	(1) This was one of the 14 constellations invented by Lacaille during his stay at the Cape of Good Hope in 1751–2.

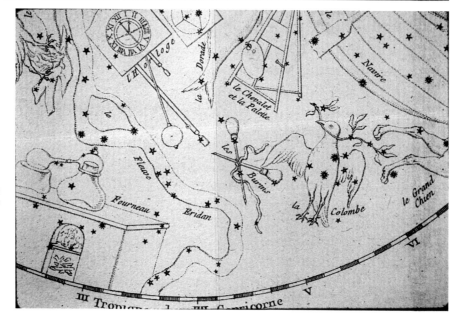

Caelum (labeled 'les Burins' on this map) Lacaille, Nicolas Louis de. Planisphere contenant les Constellations Celestes, found in Mémoires Académie Royale des Sciences, Paris, 1752 (published in 1756). This constellation was invented by Lacaille and the photo shows its first appearance on any star map.

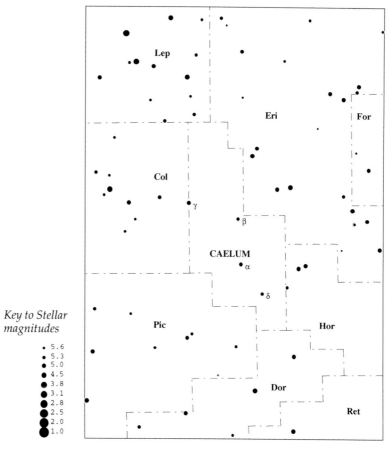

Key to Stellar magnitudes

155

Camelopardalis

Meaning:	The Giraffe
Pronunciation:	kam uh low par' dah liss
Abbreviation:	Cam
Possessive form:	Camelopardalis (kam uh low par' dah liss)
Asterisms:	none

Bordering constellations: Auriga, Cassiopeia, Cepheus, Draco, Lynx, Perseus, Ursa Major, Ursa Minor

Overall brightness:	5.946 (58)
Central point:	RA = 8h48m Dec. = +69°

Directional extremes: N = +85°   S = +53°   E = 14h25m   W = 3h11m

Messier objects:	none
Meteor showers:	none

Midnight culmination date: 23 Dec

Bright stars:	none
Named stars:	none
Near stars:	LFT 849 (41), LFT 445 (179)
Size:	756.83 square degrees (1.835% of the sky)
Rank in size:	18

Solar conjunction date: 4 Aug

Visibility:	completely visible from latitudes: N of –5°
	completely invisible from latitudes: S of –37°
Visible stars:	(number of stars brighter than magnitude 5.5): 45

Interesting facts: (1) This constellation first appeared in 1613, on a celestial globe designed by the Dutch theologian Petrus Plancius.

*Camelopardalis
Rost, Johann
Leonhard.* Atlas
Portatilis Coelestis,
Nuremburg, 1723.

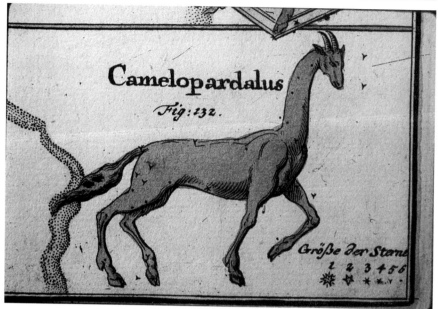

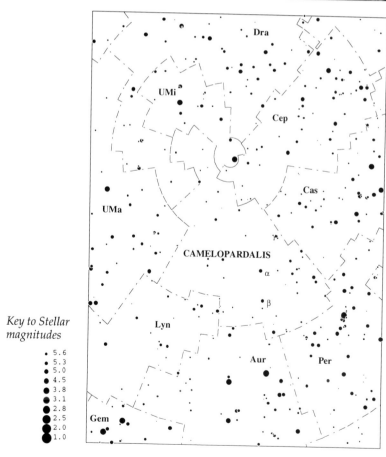

Key to Stellar
magnitudes

5.6
5.3
5.0
4.5
3.8
3.1
2.8
2.5
2.0
1.0

157

Cancer

Meaning:	The Crab
Pronunciation:	kan' sir
Abbreviation:	Cnc
Possessive form:	Cancri (kan' kree)
Asterisms:	The Asses and the Manger

Bordering constellations: Canis Minor, Gemini, Hydra, Leo, Lynx

Overall brightness:	4.545 (78)
Central point:	RA = 8h36m Dec. = +20°

Directional extremes: N = +33° S = +7° E = 9h19m W = 7h53m

Messier objects:	M44, M67
Meteor showers:	δ Cancrids (16 Jan)

Midnight culmination date: 30 Jan

Bright stars:	none
Named stars:	Acubens (α), Asellus Australis (δ), Asellus Boraelis (γ), Tarf (β), Tegmen (ζ)
Near stars:	Ross 619 (78), LP 425-140 (90)
Size:	505.87 square degrees (1.226% of the sky)
Rank in size:	31

Solar conjunction date: 1 Aug

Visibility:	completely visible from latitudes: N of –57°
	completely invisible from latitudes: S of –83°
Visible stars:	(number of stars brighter than magnitude 5.5): 23

Non-traditional 'mythology': Some see an inverted letter 'Y' composed of the brighter members of this constellation, but since these stars are quite faint to the unaided eye this figure has never attained popular acceptance.

Interesting facts: (1) The most interesting object easily visible in Cancer is M44. This is known as the 'Praesepe,' or the 'Beehive' cluster. It was first resolved into stars by Galileo, and he wrote about this in his work *Sidereus Nuncius*.

(2) The term 'Tropic of Cancer' originated several thousand years ago when the position of the summer solstice was within the boundaries of Cancer. This was the beginning of summer for the northern hemisphere and marked the day when the Sun stood highest in the sky.

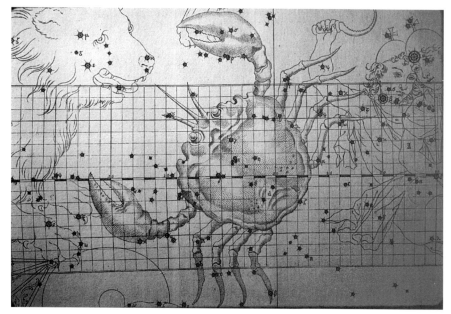

Cancer
Bevis, John.
Uranographia
Britannica, *London,*
1745.

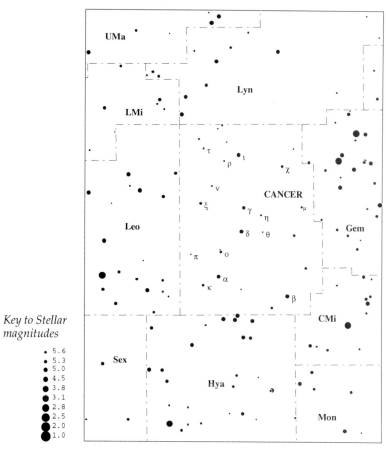

Key to Stellar
magnitudes

Canes Venatici

Meaning: The Hunting Dogs

Pronunciation: kay' neez ven ah tee' see

Abbreviation: CVn

Possessive form: Canum Venaticorum (kay' num ven at ih kor' um)

Asterisms: The Diamond (of Virgo)

Bordering constellations: Boötes, Coma Berenices, Ursa Major

Overall brightness: 3.224 (84)

Central point: RA = 13h04m   Dec. = +40.5°

Directional extremes: N = +53°   S = +28°   E = 14h05m   W = 12h04m

Messier objects: M3, M51, M63, M94, M106

Meteor showers: none

Midnight culmination date: 7 Apr

Bright stars: α^2 (150)

Named stars: Asterion (β), Chara (β), Cor Caroli (α), La Superba (Y)

Near stars: β CVn (160), BD+36°2393 (181), BD+46°1889 (196), BD+35°2436 A-B (199)

Size: 465.19 square degrees (1.128% of the sky)

Rank in size: 38

Solar conjunction date: 8 Oct

Visibility: completely visible from latitudes: N of −37°
completely invisible from latitudes: S of −62°

Visible stars: (number of stars brighter than magnitude 5.5):   15

Interesting facts: (1) One of seven constellations still in use invented by Johannes Hevelius. In 1690, this group was included in a star atlas which accompanied his stellar catalog.

(2) One-third of the way from β CVn to ζ UMa (the bend of the handle of the Big Dipper) is the unusual star Y CVn, one of the reddest stars known. Because of its strange (and apparently wonderful) spectrum, Father Secchi dubbed it 'La Superba.'

(3) One of the most beautiful spiral galaxies in the sky may be found within the confines of Canes Venatici. This is the 'Whirlpool Galaxy,' number 51 on Messier's list. In 1845, Lord Rosse saw the spiral structure in this object, making it the first galaxy to be observed with a definite spiral form.

Canes Venatici
Hevelius, Johannes.
Firmamentum
Sobiescianum, sive
Uranographia,
totum Coelum
Stellatum, *Gdansk,*
1690. This is the first
appearance of this
constellation on any
star map.

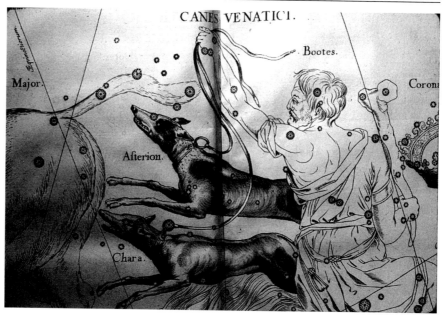

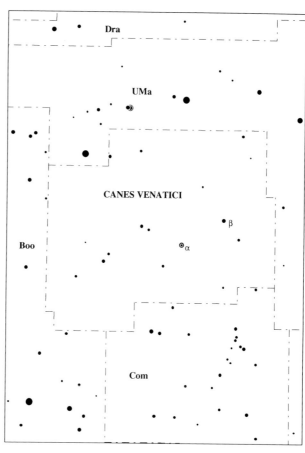

Key to Stellar
magnitudes

- 5.6
- 5.3
- 5.0
- 4.5
- 3.8
- 3.1
- 2.8
- 2.5
- 2.0
- 1.0

161

Canis Major

Meaning:	The Greater Dog
Pronunciation:	kay′ niss may′ jor
Abbreviation:	CMa
Possessive form:	Canis Majoris (kay′ niss muh jor′ iss)
Asterisms:	The Heavenly G, The Winter Octagon, The Winter Oval, The Winter Triangle

Bordering constellations: Columba, Lepus, Monoceros, Puppis

Overall brightness:	14.733 (6)
Central point:	RA = 6h47m Dec. = −22°

Directional extremes: N = −11°   S = −33°   E = 7h26m   W = 6h09m

Messier objects:	M41
Meteor showers:	none

Midnight culmination date: 2 Jan

Bright stars:	α (1), ε (22), δ (35), β (45), η (86), ζ (173), o₂ (174)
Named stars:	Adhara (ε), Aludra (η), Canicula (α), Furud (ζ), Mirzam (β), Muliphain (γ), Sirius (α), Wezen (δ)
Near stars:	α CMa A-B (6)
Size:	380.11 square degrees (0.921% of the sky)
Rank in size:	43

Solar conjunction date: 4 Jul

Visibility:	completely visible from latitudes: S of +57°
	completely invisible from latitudes: N of +79°
Visible stars:	(number of stars brighter than magnitude 5.5): 56

Interesting facts: (1) Sirius (α CMa) has the brightest apparent magnitude of any star in the nighttime sky. It appears four times as bright as Vega (α Lyr) and 25 times brighter than Polaris (α UMi). Its absolute magnitude is a respectable +0.7, making it 36 times as luminous as the sun.

(2) Sirius is a double star. Its companion, Sirius B, also known as the 'Pup,' was the first white dwarf star discovered. It was seen by Alvan Clark in 1862, while he was testing a new telescope objective lens.

(3) The heliacal rising of Sirius (the first appearance of the star in the eastern morning sky) was the major celestial occurrence in ancient Egypt. This event heralded to the annual flooding of the Nile River, which was important because it deposited a rich layer of silt over a wide area.

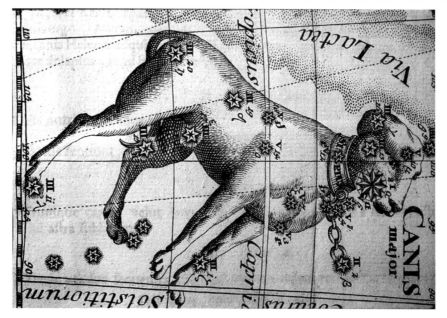

Canis Major
Thomas, Corbinianus.
Mercurii
Philosophici
Firmanentum
Firmianum,
Frankfurt and Leipzig,
1730.

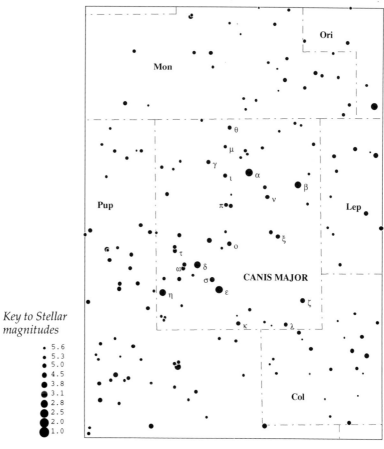

Canis Minor

Meaning:	The Lesser Dog
Pronunciation:	kay' niss my' nor
Abbreviation:	CMi
Possessive form:	Canis Minoris (kay' niss muh nor' iss)
Asterisms:	The Heavenly G, The Winter Octagon, The Winter Oval, The Winter Triangle

Bordering constellations: Cancer, Gemini, Hydra, Monoceros

Overall brightness: 7.089 (44)

Central point: RA = 7h36m   Dec. = +6.5°

Directional extremes: N = +13°   S = 00°   E = 8h09m   W = 7h04m

Messier objects: none

Meteor showers: none

Midnight culmination date: 14 Jan

Bright stars: α (8), β (149)

Named stars: Gomeisa (β), Procyon (α)

Near stars: α CMi A-B (15), BD+5°1668 (20), YZ CMi (64)

Size: 183.37 square degrees (0.445% of the sky)

Rank in size: 71

Solar conjunction date: 16 Jul

Visibility: completely visible from latitudes: N of −70
portions visible worldwide

Visible stars: (number of stars brighter than magnitude 5.5): 13

Interesting facts: (1) The name 'Procyon' (α CMi) is Greek and means 'before the Dog.' This refers to the fact that, in mid-northern latitudes, this star rises slightly before brilliant Sirius, (α CMa). Procyon was, therefore, viewed as a herald to the Dog Star.

(2) Procyon has a white dwarf companion, just like Sirius. It was first seen in 1896. This small star, known as 'Procyon B,' is about twice the size of the Earth, although it has more than half the mass of the Sun.

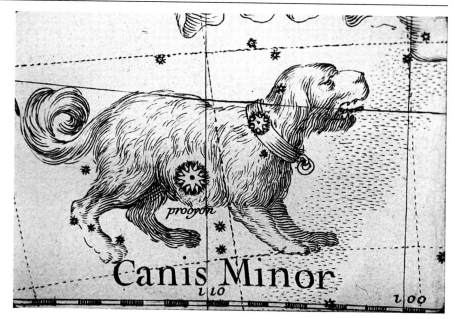

Canis Minor Pardies, Ignace Gaston. Globi Coelestis in Tabulas Planas Redacti Descriptio, *Opus posthumus, Paris, 1674.*

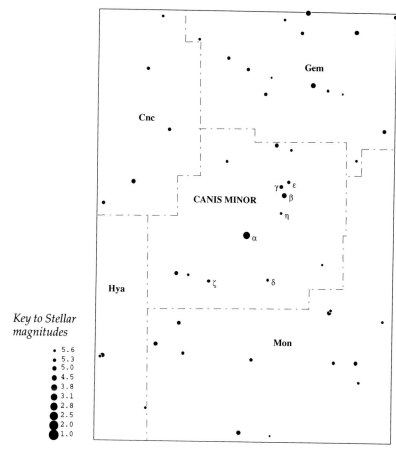

Key to Stellar magnitudes

. 5.6
. 5.3
. 5.0
. 4.5
. 3.8
. 3.1
. 2.8
. 2.5
. 2.0
. 1.0

165

Capricornus

Meaning:	The Sea Goat
Pronunciation:	kap rih kor′ nus
Abbreviation:	Cap
Possessive form:	Capricorni (kap rih corn′ ee)
Asterisms:	none

Bordering constellations: Aquarius, Aquila, Microscopium, Piscis Austrinus, Sagittarius

Overall brightness:	7.489 (36)
Central point:	RA = 21h00m Dec. = –18°

Directional extremes: N = –8°   S = –28°   E = 21h57m   W = 20h04m

Messier objects:	M30
Meteor showers:	Capricornids (22 Jul)
	α Capricornids (30 Jul)

Midnight culmination date: 8 Aug

Bright stars:	δ (142), β (185)
Named stars:	Algedi (α), Alshat (ν), Dabih (β), Deneb Algedi (δ), Giedi (α), Gredi (α), Nashira (γ), Prima Giedi (α)
Near stars:	Wolf 922 (96), LFT 1535 (125)
Size:	413.95 square degrees (1.003% of the sky)
Rank in size:	40

Solar conjunction date: 5 Feb

Visibility:	completely visible from latitudes: S of +62°
	completely invisible from latitudes: N of +82°
Visible stars:	(number of stars brighter than magnitude 5.5): 31

Non-traditional 'mythology': This constellation is often called the "smile in the sky." This smile is composed of the stars δ, γ, ι, θ, α, β, π, ψ, ω, 24, ζ, and ε of this constellation. However, a lady's high-heeled shoe may also be found among these stars. Begin with α as the toe of the shoe, then move to β, π, and then over to χ; then proceed down to ψ, then to ω, 24, ζ, δ, γ, ι, and θ.

Interesting facts: (1) Of the 12 traditional constellations of the zodiac, Capricornus is the smallest.

(2) According to Flammarion, Chinese astronomers observed five planets in conjunction in this constellation in 2449 BC.

(3) Over 2000 years ago, the position of the Sun at the December solstice lay within the confines of this constellation. At that point, the Sun was at declination –23.5°, or 23.5° below the celestial equator. The corresponding latitude on Earth (the southernmost point where the Sun could be directly overhead at noon) thus was named the 'Tropic of Capricorn.' It still retains this designation today, even though the Earth's precessional motion has moved the point of the December solstice into neighboring Sagittarius.

*Capricornus
Semler, Christoph.
Coelum Stellatum in
quo asterisimi...,
Halle, 1731.*

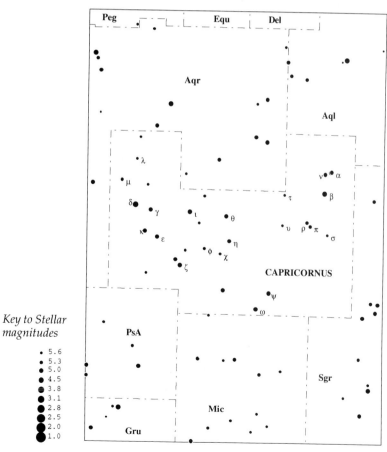

*Key to Stellar
magnitudes*

167

Carina

Meaning:	The Keel (of Argo Navis)
Pronunciation:	kuh ree' nuh
Abbreviation:	Car
Possessive form:	Carinae (kar ee' nye)
Asterisms:	The False Cross

Bordering constellations: Centaurus, Chamaeleon, Musca, Pictor, Puppis, Vela, Volans

Overall brightness:	15.581 (3)
Central point:	RA = 8h40m Dec. = −63°

Directional extremes: N= −51°   S= −75°   E= 11h18m   W= 6h02m

Messier objects:	none
Meteor showers:	none

Midnight culmination date: 31 Jan

Bright stars:	α (2), β (28), ε (37), ι (68), θ (120), υ (159)
Named stars:	Asmidiske (ι), Avior (ε), Canopus (α), Miaplacidus (β), Scutulum (ι), Tureis (ι)
Near stars:	LFT 643 (172)
Size:	494.18 square degrees (1.198% of the sky)
Rank in size:	34

Solar conjunction date: 2 Aug

Visibility:	completely visible from latitudes: S of +15°
	completely invisible from latitudes: N of +39°
Visible stars:	(number of stars brighter than magnitude 5.5): 77

Interesting facts:

(1) This is one of three constellations formed out of the immense star group known as 'Argo Navis.' In legend the Argo was the ship Jason and the Argonauts used to search for the Golden Fleece. Carina represents the keel of the ship. The other two constellations are Puppis and Vela.

(2) Canopus (α Car) is the second brightest of the nighttime stars, outshone only by Sirius (α CMa). Sirius appears to the eye almost exactly twice as bright as Canopus.

(3) Although Canopus is now the second brightest of all stars, there was a time, in 1843, when the star η Car was brighter. η Car is a nova-like variable star whose outbursts during the eighteenth and nineteenth centuries represent a fluctuation in brightness of over 4000 times! η Car is located within a diffuse interstellar cloud known as NGC 3372, a portion of which is often referred to as the 'Keyhole Nebula' due to its shape.

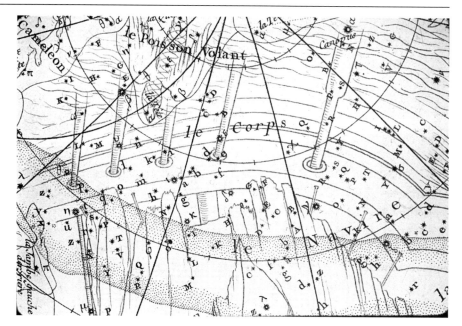

Carina (labeled 'le Corps' on this map) Vaugondy, Robert de. Hémisphère Céleste Antarctique..., *Paris, 1764. This constellation was invented by Lacaille and included in his star catalog, but not pictured on any of his star charts. This photo from Robert de Vaugondy's map, therefore, shows the first appearance of Carina as a separate constellation.*

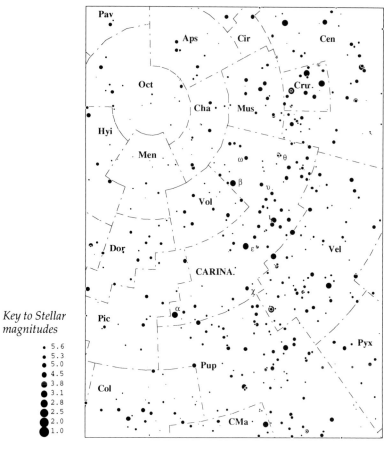

Key to Stellar magnitudes

169

Cassiopeia

Meaning:	The Queen of Ethiopia
Pronunciation:	kass ee oh pee' uh
Abbreviation:	Cas
Possessive form:	Cassiopeiae (kass ee oh pee' eye)
Asterisms:	The Three Guides

Bordering constellations: Andromeda, Camelopardalis, Cepheus, Lacerta, Perseus

Overall brightness:	8.523 (30)
Central point:	RA = 1h16m Dec. = +62°
Directional extremes:	N = +78° S = +46° E = 3h36m W = 22h56
Messier objects:	M52, M103
Meteor showers:	none

Midnight culmination date: 9 Oct

Bright stars:	α (64), β (70), γ (88), δ (105)
Named stars:	Achird (η), Caph (β), Cih (γ), Ksora (δ), Navi (ε), Ruchbah (δ), Schedar (α), Segin (ε), Tsih (γ)
Near stars:	η Cas A-B (55), BD+56°2966 (82), μ Cas (112), Ross 318 (138), BD+63°238 (140), V388 Cas (146), Wolf 46 (155)
Size:	598.41 square degrees (1.451% of the sky)
Rank in size:	25

Solar conjunction date: 11 Apr

Visibility:	completely visible from latitudes: N of –12°
	completely invisible from latitudes: S of –44°
Visible stars:	(number of stars brighter than magnitude 5.5): 51

Interesting facts: (1) The asterism The Three Guides, formed by the stars β Cas, α And, and γ Peg mark the equinoctial colure. This is the great circle which intersects both celestial poles and both equinoxes.

(2) B Cas, also known as 'Tycho's Star,' represents the supernova of 1572, one of only four observed supernovae known to have originated in our galaxy. The others are the novae of 1006 in Lupus, 1054 in Taurus (the event which produced the Crab Nebula), and 1604 in Ophiuchus (now known as 'Kepler's Star'). Tycho's Star was visible to the unaided eye for over 16 months and at its brightest this object could be easily detected in full daylight.

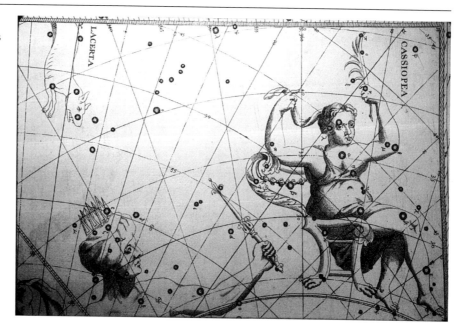

Cassiopeia
Flamsteed, John. Atlas
Coelestis, *London,*
1729.

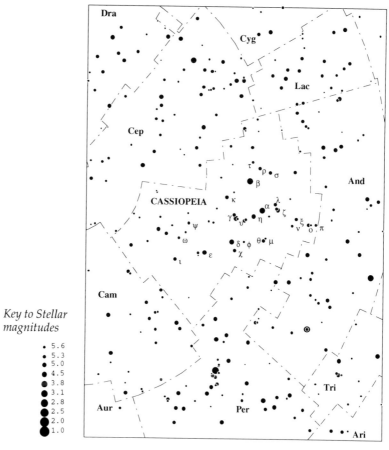

Key to Stellar
magnitudes

171

Centaurus

Meaning:	The Centaur
Pronunciation:	sen tor' us
Abbreviation:	Cen
Possessive form:	Centauri (sen tor' ee)
Asterisms:	The Southern Pointers

Bordering constellations: Antlia, Carina, Circinus, Crux, Hydra, Lupus, Musca, Vela

Overall brightness: 9.525 (25)

Central point: RA = 13h01m Dec. = –47.5°

Directional extremes: N = –30° S = –65° E = 14h59m W = 11h03m

Messier objects: none

Meteor showers: α Centaurids (8 Feb)

Midnight culmination date: 30 Mar

Bright stars: α (3), β (11), θ (54), γ (60), ε (73), η (75), ζ (92), δ (98), ι (118), μ (177), λ (193), κ (194)

Named stars: Agena (β), Hadar (β), Menkent (θ), Proxima Centauri (α Cen C), Rigil Kentaurus (α), Toliman (α)

Near stars: α Cen C (1), α Cen A-B (2), LFT 930 (122), LFT 839 (123), LFT 1088 (173)

Size: 1060.42 square degrees (2.571% of the sky)

Rank in size: 9

Solar conjunction date: 7 Oct

Visibility: completely visible from latitudes: S of +25°
completely invisible from latitudes: N of +60

Visible stars: (number of stars brighter than magnitude 5.5): 101

Interesting facts: (1) The nearest star system to our own is that of α Cen, or Rigil Kentaurus. Although this star has a proper name, it is almost universally referred to as Alpha Centauri. The distance to α Cen is 4.39 light years. Alpha Centauri is a triple star system, the two brighter components making a wonderful double star to observe in even small instruments. The third star in this group is commonly referred to as Proxima Centauri, as it lies slightly nearer to the Earth – about 0.1 light years – than the aforementioned pair. Visually, α Cen is the third brightest of all nighttime stars, shining with an apparent magnitude of –0.27.

(2) As α Cen has a large proper motion around the year 6000 it will have moved near β Cen and they will become a fabulous double star.

(3) The finest globular cluster in the sky is ω Cen, the Omega Centauri cluster. The designation of this object as a star comes from ancient times, with the Greek letter omega being given it by Johannes Bayer in 1603, in his *Uranometria* star atlas. Today, Omega Centauri is also referred to by its official designation: NGC 5139. The total visual magnitude of ω Cen is a relatively bright 3.7, but its southerly declination (–47°) makes it a difficult object to view at latitudes above 35° N.

*Centaurus
Cellarius, Andreas.*
Harmonia
Macrocosmica sev
Atlas Universalis et
Novus, *Amsterdam,
1661.*

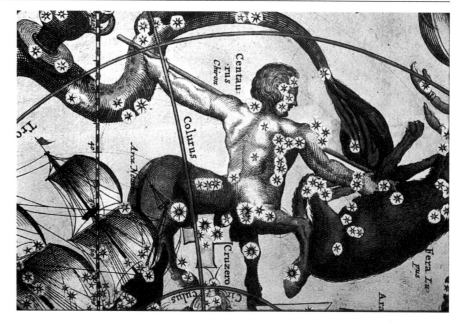

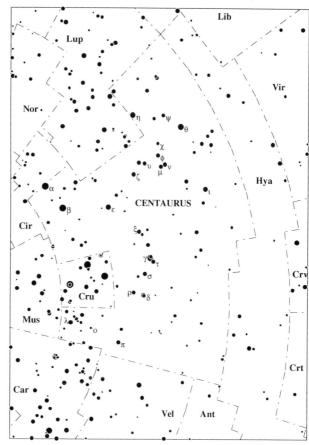

*Key to Stellar
magnitudes*

Cepheus

Meaning:	The King of Ethiopia
Pronunciation:	see' fee us
Abbreviation:	Cep
Possessive form:	Cephei (see' fee ee)
Asterisms:	none

Bordering constellations: Camelopardalis, Cassiopeia, Cygnus, Draco, Lacerta, Ursa Minor

Overall brightness:	9.697 (22)
Central point:	RA = 2h15m Dec. = +70°
Directional extremes:	N = +89° S = +51° E = 8h30m W = 20h01m
Messier objects:	none
Meteor showers:	none

Midnight culmination date: 29 Sep

Bright stars:	α (85)
Named stars:	Alderamin (α), Alfirk (β), Alrai (γ), Erakis (μ), Er Rai (γ), Kurhah (ξ)
Near stars:	Krueger 60 A-B (23), BD+61°2068 (91)
Size:	587.79 square degrees (1.425% of the sky)
Rank in size:	27

Solar conjunction date: 26 Apr

Visibility:	completely visible from latitudes: N of −1°
	completely invisible from latitudes: S of −39°
Visible stars:	(number of stars brighter than magnitude 5.5): 57

Non-traditional 'mythology': For many school children in northern latitudes, the "house" in Cepheus is required learning. This figure is composed of the stars γ, β, α, δ, and ι.

Interesting facts: (1) μ Cep is often called 'Herschel's Garnet Star' due to the fact that it is one of the reddest stars visible. μ Cep is a variable star, changing its apparent magnitude from 3.6 to 5.1 on an irregular basis.

(2) δ Cep is the original Cepheid variable. This star's changes in brightness were first detected in 1784, by John Goodricke. The variability of the light of δ Cep is caused by changes in the radius of the star – pulsations of the outer layers of its atmosphere. Henrietta Leavitt, working at the Harvard College Observatory from 1908–12, discovered that there was a relationship between the period of the pulsations of this type of star and its luminosity. This 'period–luminosity' relationship has become one of the 'yardsticks' by which distances to stars, clusters, and even nearby galaxies are measured.

(3) Recently (1993), within the boundaries of this constellation, astronomers have discovered a star moving through space extremely rapidly. The 'wake' this object is leaving behind as it moves through the galaxy has the shape of a guitar. It has come to be known as the 'Guitar Nebula.'

Cepheus
Hyginus. Poeticon
Astronomicon,
Venice, 1485.

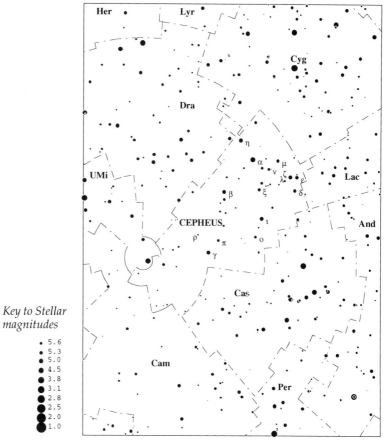

*Key to Stellar
magnitudes*

175

Cetus

Meaning:	The Sea Monster (Whale)
Pronunciation:	see' tus
Abbreviation:	Cet
Possessive form:	Ceti (set' ee)
Asterisms:	The Head

Bordering constellations: Aquarius, Aries, Eridanus, Fornax, Pisces, Sculptor, Taurus

Overall brightness:	4.710 (77)
Central point:	RA= 1h38m Dec. = −7.5°

Directional extremes: N = +10°   S = −25°   E = 3h21m   W = 23h55m

Messier objects:	M77
Meteor showers:	none

Midnight culmination date: 15 Oct

Bright stars:	β (50), α (91)
Named stars:	Baten Kaitos (ζ), Deneb al Schemali (ι), Deneb Kaitos (β), Diphda (β), Kaffaljidhmah (γ), Menkar (α), Mira (o)
Near stars:	UV Cet A-B (7), τ Cet (18), BD+6°398 A-B (85), BD–18°359 (141), χ Cet (163), Wolf 124 (193)
Size:	1231.41 square degrees (2.985% of the sky)
Rank in size:	4

Solar conjunction date: 17 Apr

Visibility:	completely visible from latitudes: S of +65°
	portions visible worldwide
Visible stars:	(number of stars brighter than magnitude 5.5): 58

Non-traditional 'mythology': This constellation has occasionally been referred to as the 'easy chair,' due to the similarity between the layout of its stars and a reclining chair.

Interesting facts: (1) o Cet was the first variable star discovered and remains one of the most famous of all such stars in the sky. It is a red giant whose brightness at minimum is about 8th to 10th magnitude. At maximum, about 331 days later, it is usually as bright as 3rd or 4th magnitude. Once, in 1779, its brightness rivaled that of Aldebaran (α Tau) and as recently as 1969 its apparent magnitude was measured at 2.1. The variability of o Cet was noted by David Fabricius in 1596. Later, Hevelius gave it the name 'Mira,' which translated means 'the wonderful.' It is now known that Mira belongs to a class of stars known as pulsating variables, stars which vary in brightness due to changes in their size.

(2) τ Cet has become the focus of much attention in recent years. It is an individual solar-type star lying at a distance of only 11.68 light years from Earth. Only one such star is closer – ε Ind, which is 11.2 light years away. These stars have been studied intensely by astronomers searching for planetary systems which may harbor intelligent life.

*Cetus (labeled 'Balena'
on this map)
Coronelli, Vincenzo.*
Epitome
Cosmografica...,
Cologne, 1693.

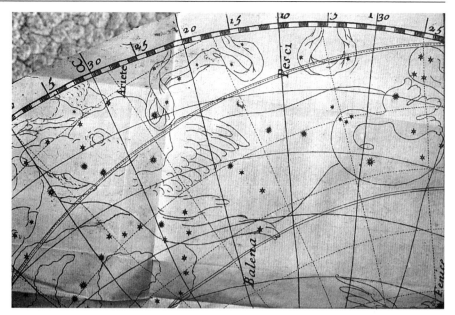

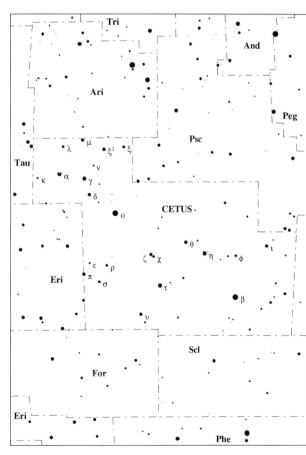

*Key to Stellar
magnitudes*

5.6
5.3
5.0
4.5
3.8
3.1
2.8
2.5
2.0
1.0

177

Chamaeleon

Meaning:	The Chameleon
Pronunciation:	kuh meel' ee un
Abbreviation:	Cha
Possessive form:	Chamaeleontis (kuh meel ee on' tiss)
Asterisms:	none

Bordering constellations: Apus, Carina, Mensa, Musca, Octans, Volans

Overall brightness:	9.879 (20)
Central point:	RA = 10h40m Dec. = –79°
Directional extremes:	N = –75° S = –83° E = 13h48m W = 7h32m
Messier objects:	none
Meteor showers:	none

Midnight culmination date: 1 Mar

Bright stars:	none
Named stars:	none
Near stars:	none
Size:	131.59 square degrees (0.319% of the sky)
Rank in size:	79

Solar conjunction date: 1 Sep

Visibility:	completely visible from latitudes: S of +7°
	completely invisible from latitudes: N of +15°
Visible stars:	(number of stars brighter than magnitude 5.5): 13
Interesting facts:	(1) This is one of 11 constellations invented by Pieter Dirksz Keyser and Frederick de Houtman, during the years 1595–97.

Chamaeleon
Bayer, Johann.
Uranometria,
Augsburg, 1603. This
constellation was
invented by de
Houtman and Keyser
in 1596. It was first
illustrated on a globe
by Plancius, which has
not survived. This
photo from Bayer's
map, therefore, shows
the earliest existing
picture of this
constellation.

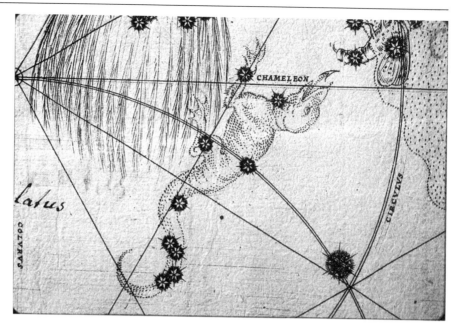

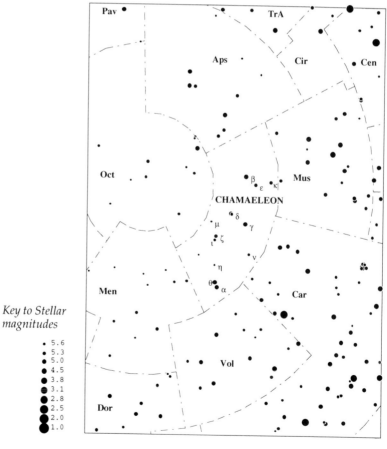

Key to Stellar
magnitudes

- 5.6
- 5.3
- 5.0
- 4.5
- 3.8
- 3.1
- 2.8
- 2.5
- 2.0
- 1.0

Circinus

Meaning:	The Compasses
Pronunciation:	sir sin' us
Abbreviation:	Cir
Possessive form:	Circini (sir sin' ee)
Asterisms:	none
Bordering constellations:	Apus, Centaurus, Lupus, Musca, Norma, Triangulum Australe
Overall brightness:	10.712 (16)
Central point:	RA = 14h30m Dec. = –62°
Directional extremes:	N = –54° S = –70° E = 15h26m W = 13h35m
Messier objects:	none
Meteor showers:	none
Midnight culmination date:	30 Apr
Bright stars:	none
Named stars:	none
Near stars:	none
Size:	93.35 square degrees (0.226% of the sky)
Rank in size:	85
Solar conjunction date:	29 Oct
Visibility:	completely visible from latitudes: S of +30°
	completely invisible from latitudes: N of +36°
Visible stars:	(number of stars brighter than magnitude 5.5): 10
Interesting facts:	(1) This was one of the 14 constellations invented by Lacaille during his stay at the Cape of Good Hope in 1751–2.

Circinus (labeled 'le Compas' on this map) Lacaille, Nicolas Louis de. Planisphere contenant les Constellations Celestes, found in Mémoires Académie Royale des Sciences, *Paris, 1752 (published in 1756). This constellation was invented by Lacaille and the photo shows its first appearance on any star map.*

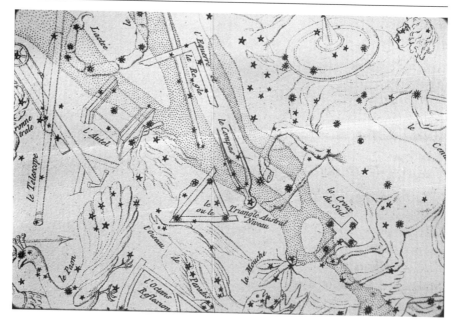

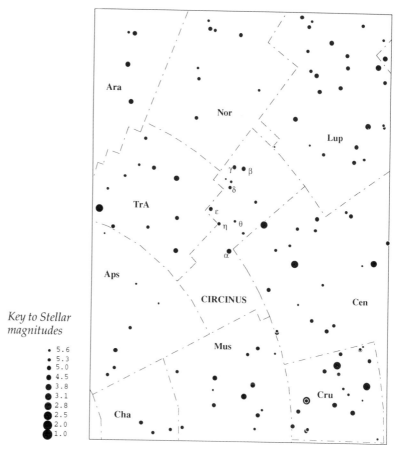

Key to Stellar magnitudes

- 5.6
- 5.3
- 5.0
- 4.5
- 3.8
- 3.1
- 2.8
- 2.5
- 2.0
- 1.0

Columba

Meaning:	Noah's Dove
Pronunciation:	kol um' buh
Abbreviation:	Col
Possessive form:	Columbae (kol um' bye)
Asterisms:	none

Bordering constellations: Caelum, Canis Major, Lepus, Pictor, Puppis

Overall brightness:	8.883 (29)
Central point:	RA = 5h45m Dec. = –35°
Directional extremes:	N = –27° S = –43° E = 6h28m W = 5h03m
Messier objects:	none
Meteor showers:	none

Midnight culmination date: 18 Dec

Bright stars:	α (102), β (190)
Named stars:	Phakt (α), Wasn (β)
Near stars:	none
Size:	270.18 square degrees (0.655% of the sky)
Rank in size:	54

Solar conjunction date: 18 Jun

Visibility:	completely visible from latitudes: S of +47°
	completely invisible from latitudes: N of +63°
Visible stars:	(number of stars brighter than magnitude 5.5): 24

Interesting facts: (1) Columba is the only surviving constellation named after an object in the Bible. Columba represents the dove which Noah sent out to test whether the waters from the Great Flood had receded. (Genesis, chapter 8, verses 8–12). This constellation first appeared in 1592, on a celestial map designed by the Dutch theologian Petrus Plancius.

(2) The so-called solar antapex is located in this constellation. This is the direction in space away from which our Sun seems to be heading. It is the point opposite the solar apex, which lies within the constellation of Hercules. The approximate coordinates of the solar antapex are RA = 6 h, Dec. = –34°. (The Sun's motion is relative to the stars in the local neighborhood and is determined by studies of parallaxes, proper motions, and radial velocities of nearby stars.)

Columba
Schiller, Julius.
Coelum Stellatum
Christianum,
Augsburg, 1627.

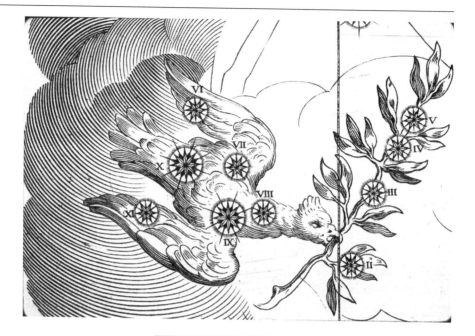

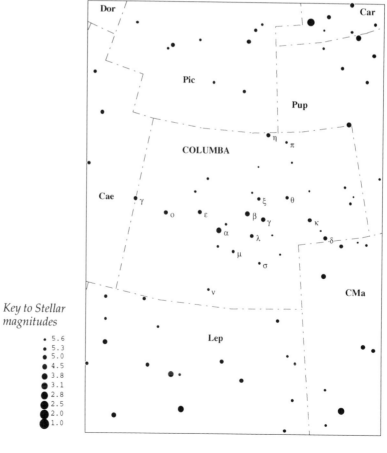

*Key to Stellar
magnitudes*

- . 5.6
- . 5.3
- . 5.0
- 4.5
- 3.8
- 3.1
- 2.8
- 2.5
- 2.0
- 1.0

183

Coma Berenices

Meaning:	The Hair of Berenice
Pronunciation:	koe' muh bear uh nye' seez
Abbreviation:	Com
Possessive form:	Comae Berenices (koe' my ber uh nye' seez)
Asterisms:	none
Bordering constellations:	Boötes, Canes Venatici, Leo, Ursa Major, Virgo
Overall brightness:	5.951 (57)
Central point:	RA = 12h45m Dec. = +23.5°
Directional extremes:	N = +34° S = +13° E = 13h33m W = 11h57m
Messier objects:	M53, M64, M85, M88, M91, M98, M99, M100
Meteor showers:	Coma Berenicids (10 Jan)
Midnight culmination date:	2 Apr
Bright stars:	none
Named stars:	Diadem (α)
Near stars:	β Com (124)
Size:	386.47 square degrees (0.937% of the sky)
Rank in size:	42
Solar conjunction date:	3 Oct
Visibility:	completely visible from latitudes: N of −56°
	completely invisible from latitudes: S of −77°
Visible stars:	(number of stars brighter than magnitude 5.5): 23

Interesting facts:
(1) This constellation was created by Gerard Mercator on a globe designed by him in 1551.

(2) The north galactic pole lies within the constellation of Coma Berenices. Its approximate 1950.0 coordinates are RA = 12h49m Dec. = +27.4°.

(3) One of the most famous galaxies in the sky, and one whose photograph is often reproduced to show a classic 'edge-on' spiral galaxy is NGC 4565. This object lies slightly more than 3° to the southeast of γ Com. In a dark sky, even small instruments will show this 10th magnitude object to be a slender filament of light. Only the largest telescopes reveal its true character and beauty, however. NGC 4565 lies at a distance of approximately 20 million light years.

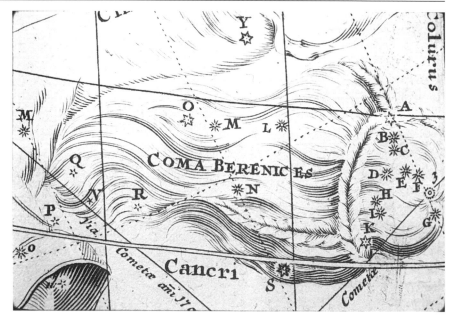

Coma Berenices Doppelmayr, Johann Gabriel. Atlas Coelestis, Nuremburg, 1742.

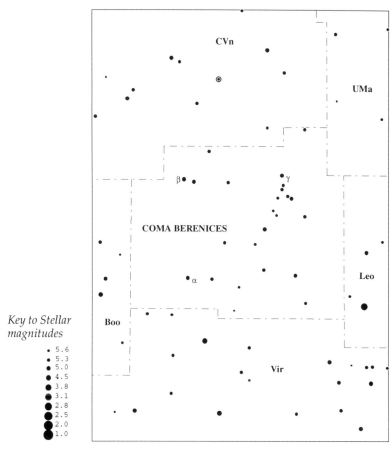

Key to Stellar magnitudes

- 5.6
- 5.3
- 5.0
- 4.5
- 3.8
- 3.1
- 2.8
- 2.5
- 2.0
- 1.0

185

Corona Australis

Meaning:	The Southern Crown
Pronunciation:	kor oh' nuh os tral' iss
Abbreviation:	CrA
Possessive form:	Coronae Australis (kor oh' nye os tral' iss)
Asterisms:	none

Bordering constellations: Ara, Sagittarius, Scorpius, Telescopium

Overall brightness:	16.446 (2)
Central point:	RA = 18h35m Dec. = −41.5°

Directional extremes: N = −37°   S = −46°   E = 19h15m   W = 17h55m

Messier objects:	none
Meteor showers:	Corona Australids (16 Mar)

Midnight culmination date: 30 Jun

Bright stars:	none
Named stars:	none
Near stars:	none
Size:	127.69 square degrees (0.310%)
Rank in size:	80

Solar conjunction date: 31 Dec

Visibility:	completely visible from latitudes: S of +44°
	completely invisible from latitudes: N of +53°
Visible stars:	(number of stars brighter than magnitude 5.5): 21

Non-traditional 'mythology': If the brightest stars of Sagittarius form a 'teapot,' then the stars of this constellation, notably γ, α, β, δ, ζ, η, and θ, form a slice of lemon near the teapot.

Interesting facts: (1) γ CrA is an interesting double star. Both components are main-sequence stars of spectral type F8, and they are almost exactly the same brightness, with visual magnitudes of 4.84 and 5.08. In the 1881 revision of Smyth's *Cycle of Celestial Objects*, Chambers quotes Sir John Herschel as calling this star 'superb.'

Corona Australis
Burritt, Elijah H.
Atlas Designed to
Illustrate the
Geography of the
Heavens, *Hartford,*
1835.

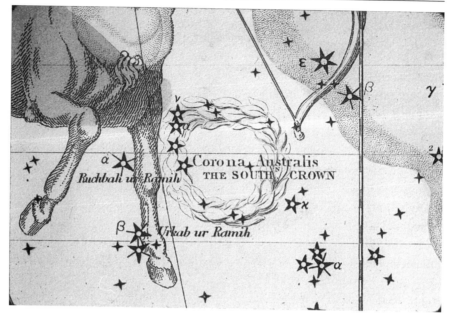

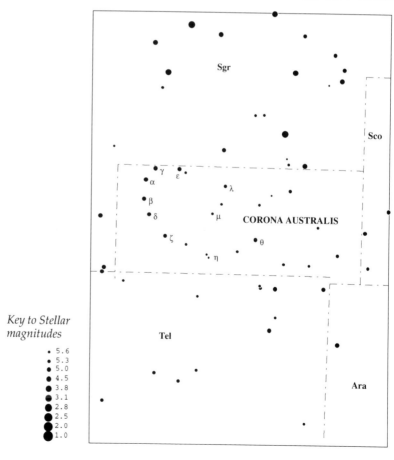

Key to Stellar
magnitudes

· 5.6
· 5.3
· 5.0
● 4.5
● 3.8
● 3.1
● 2.8
● 2.5
● 2.0
● 1.0

187

Corona Borealis

Meaning:	The Northern Crown
Pronunciation:	kor oh' nuh boar ee al' iss
Abbreviation:	CrB
Possessive form:	Coronae Borealis (kor oh' nye bor ee al' iss)
Asterisms:	none

Bordering constellations: Boötes, Hercules, Serpens

Overall brightness:	12.310 (11)
Central point:	RA = 15h48m Dec. = +33°

Directional extremes: N = +40°   S = +26°   E = 16h22m   W = 15h14m

Messier objects:	none
Meteor showers:	none

Midnight culmination date: 19 May

Bright stars:	α (65)
Named stars:	Alphecca (α), Gemma (α), Nusakan (β)
Near stars:	none
Size:	178.71 square degrees (0.433% of the sky)
Rank in size:	73

Solar conjunction date: 18 Nov

Visibility:	completely visible from latitudes: N of –50°
	completely invisible from latitutdes: S of –64°
Visible stars:	(number of stars brighter than magnitude 5.5): 22

Interesting facts: (1) On 12 May 1866, a nova suddenly flared up near ε CrB. It reached 2nd magnitude and remained visible to the unaided eye for eight nights. It has been given the designation T CrB, but it is more commonly known as the 'Blaze Star.' This object is the most famous example of a recurring nova. It brightened again (to 3rd magnitude) on 9 February 1946.

(2) Another interesting variable star lies within this constellation. It is known as R CrB. The normal brightness of this star is 6th magnitude, but at an irregular interval of several to many years its light output may drop to as faint as 15th magnitude. The reason for this is not well understood, however it is believed that clouds of carbon (graphite or soot) are emitted from the star, dimming its light. Then, as the material is reabsorbed, the brightness of the star returns to normal.

*Corona Borealis
Bode, Johann Elert.*
Uranographia Sive
Astrorum
Descriptio, *Berlin,
1801.*

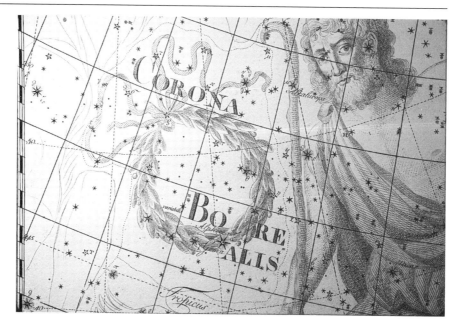

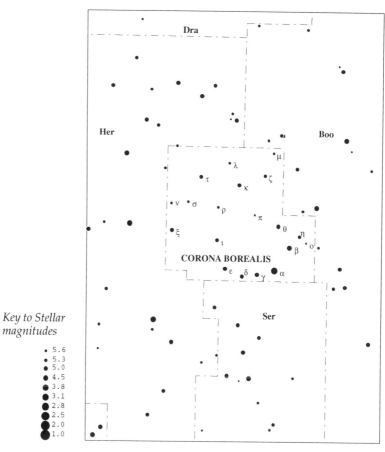

*Key to Stellar
magnitudes*

· 5.6
· 5.3
· 5.0
· 4.5
· 3.8
· 3.1
· 2.8
· 2.5
· 2.0
· 1.0

189

Corvus

Meaning:	The Crow
Pronunciation:	kor' vus
Abbreviation:	Crv
Possessive form:	Corvi (kor' vee)
Asterisms:	The Sail

Bordering constellations: Crater, Hydra, Virgo

Overall brightness:	5.985 (56)
Central point:	RA = 12h24m Dec. = −18°

Directional extremes: N = −11°   S = −25°   E = 12h54m   W = 11h54m

Messier objects:	none
Meteor showers:	Corvids (26 Jun)

Midnight culmination date: 28 Mar

Bright stars:	γ (97), β (103), δ (157), ε (168)
Named stars:	Alchiba (α), Algorab (δ), Kraz (β), Minkar (ε)
Near stars:	Ross 695 (154)
Size:	183.80 square degrees (0.446% of the sky)
Rank in size:	70

Solar conjunction date: 27 Sep

Visibility:	completely visible from latitudes: S of +65°
	completely invisible from latitudes: N of +79°
Visible stars:	(number of stars brighter than magnitude 5.5): 11

Non-traditional 'mythology': The four brightest stars of this constellation (γ, β, δ, and ε) are often referred to as a sail. Also, a set of 'mini-Pointers' is present, as a line drawn from γ through δ points to Spica (α Vir).

Interesting facts: (1) A very unusual object, NGC 4038–39, lies within the boundaries of this constellation. This is a pair of interacting (possibly, colliding) galaxies often called the 'Ring-tail Galaxy.' Its appearance on photographs taken with large instruments is that of a heart. Two streamers originating from the interaction point can also be seen. This object is classified as a peculiar galaxy.

Corvus
Flamsteed, John. Atlas
Coelestis, *London,*
1729.

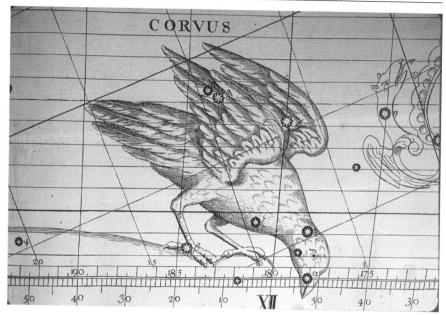

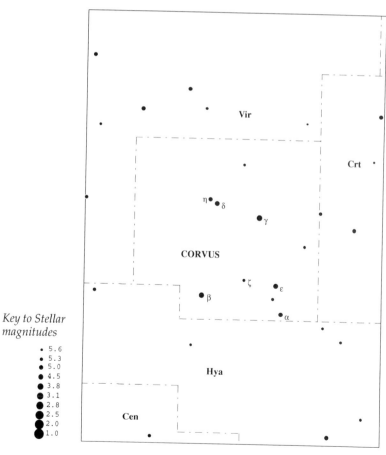

Key to Stellar
magnitudes

- 5.6
- 5.3
- 5.0
- 4.5
- 3.8
- 3.1
- 2.8
- 2.5
- 2.0
- 1.0

Crater

Meaning:	The Cup
Pronunciation:	kray′ ter
Abbreviation:	Crt
Possessive form:	Crateris (kray ter′ iss)
Asterisms:	none

Bordering constellations: Corvus, Hydra, Leo, Sextans, Virgo

Overall brightness:	3.895 (82)
Central point:	RA = 11h21m Dec. = −15.5°

Directional extremes: N = −6°   S = −25°   E = 11h54m   W = 10h48m

Messier objects:	none
Meteor showers:	none

Midnight culmination date: 12 Mar

Bright stars:	none
Named stars:	Alkes (α)
Near stars:	LFT 764 (189), LTT 4204 A-B (194)
Size:	282.40 square degrees (0.685% of the sky)
Rank in size:	53

Solar conjunction date: 11 Sep

Visibility:	completely visible from latitudes: S of 65°
	completely invisible from latitudes: N of +84°
Visible stars:	(number of stars brighter than magnitude 5.5): 11

Interesting facts: (1) As the only named star in the constellation, Alkes may have been brighter several hundred years ago when it was designated α Crt by Bayer. Today, δ Crt is the brightest star, with an apparent magnitude of 3.56. α Crt is more than a half magnitude fainter, shining at magnitude 4.08.

Crater
Doppelmayr, Johann
Gabriel. Atlas
Coelestis,
Nuremburg, 1742.

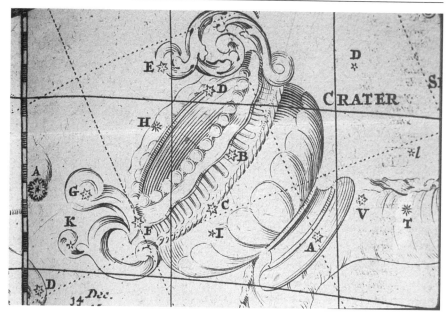

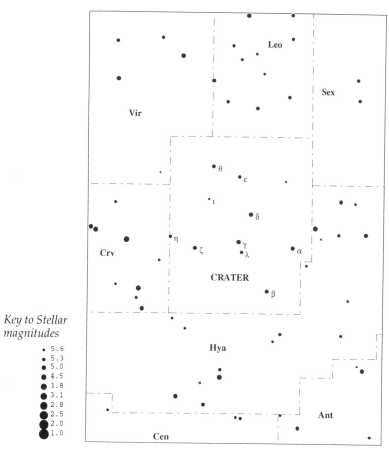

Key to Stellar
magnitudes

- 5.6
- 5.3
- 5.0
- 4.5
- 3.8
- 3.1
- 2.8
- 2.5
- 2.0
- 1.0

Crux

Meaning:	The (Southern) Cross
Pronunciation:	kruks
Abbreviation:	Cru
Possessive form:	Crucis (kroo' siss)
Asterisms:	none

Bordering constellations: Centaurus, Musca

Overall brightness:	29.218 (1)
Central point:	RA = 12h24m Dec. = −60°

Directional extremes: N = −55°   S = −65°   E = 12h55m   W = 11h53m

Messier objects:	none
Meteor showers:	none

Midnight culmination date: 28 Mar

Bright stars:	α (13), β (19) γ (24), δ (127)
Named stars:	Acrux (α), Gacrux (γ), Mimosa (β)
Near stars:	none
Size:	68.45 square degrees (0.166% of the sky)
Rank in size:	88

Solar conjunction date: 27 Sep

Visibility:	completely visible from latitudes: S of +25°
	completely invisible from latitudes: N of +35°
Visible stars:	(number of stars brighter than magnitude 5.5): 20

Interesting facts: (1) This constellation was formed by early European explorers. The first reliable reference we have about Crux comes from a letter written by the Italian navigator Amerigo Vespucci, in 1503. Crux was important to early navigators because a line drawn from γ Cru through α Cru and extended some 25° roughly points to the south celestial pole. There is no bright star near this important directional point as exists in the northern hemisphere.

(2) A famous dark nebula known as the 'Coal Sack' lies generally south of β Cru and east of α Cru, covering more than 30° of sky area. This nebula is easily seen with the unaided eye, silhouetted, as it is, against the brilliant and closely-packed stars of the southern Milky Way.

(3) One of the most beautiful of all star clusters lies within the boundaries of Crux. This is NGC 4755, known either as the 'Kappa Crucis Star Cluster' or the 'Jewel Box.' This latter title was inferred by a statement made by John Herschel comparing this stellar group to a piece of multicolored jewelry.

Crux
Bode, Johann Elert.
Uranographia Sive
Astrorum
Descriptio, *Berlin,*
1801.

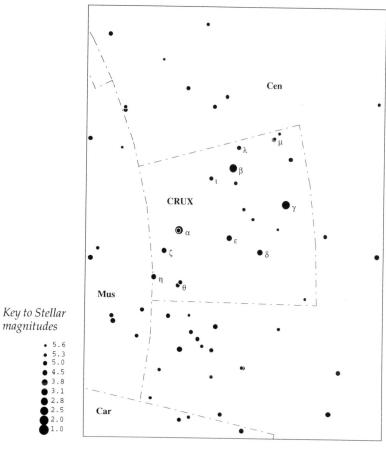

Key to Stellar
magnitudes

- 5.6
- 5.3
- 5.0
- 4.5
- 3.8
- 3.1
- 2.8
- 2.5
- 2.0
- 1.0

195

Cygnus

Meaning:	The Swan
Pronunciation:	sig′ nus
Abbreviation:	Cyg
Possessive form:	Cygni (sig′ nee)
Asterisms:	The Northern Cross, The Summer Triangle

Bordering constellations: Cepheus, Draco, Lacerta, Lyra, Pegasus, Vulpecula

Overall brightness:	9.826 (21)
Central point:	RA = 20h34m Dec. = +44.5°

Directional extremes: N = +61° S = +28° E = 22h01m W = 19h07m

Messier objects:	M29, M39
Meteor showers:	κ Cygnids (18 Aug)

Midnight culmination date: 30 Jul

Bright stars:	α (20), γ (61), ε (87), δ (141), β (184)
Named stars:	Albireo (β), Azelfafage (π′), Deneb (α), Gienah (ε), Sadr (γ)
Near stars:	61 Cyg A-B (13)
Size:	803.98 square degrees (1.949% of the sky)
Rank in size:	16

Solar conjunction date: 30 Jan

Visibility:	completely visible from latitudes: N of –29°
	completely invisible from latitudes: S of –62°
Visible stars:	(number of stars brighter than magnitude 5.5): 79

Non-traditional 'mythology': The asterism of the Northern Cross has long been recognized and its meaning is enhanced during the holiday season when, on Christmas Eve in northern latitudes, it stands nearly upright on the northwestern horizon at sunset.

Interesting facts: (1) Deneb (α Cyg) is one of the most luminous stars in the sky. It has a luminosity of about 60,000 times that of the Sun, corresponding to an absolute magnitude of –7.1.

(2) About 3° to the east of Deneb is NGC 7000, known as the 'North American Nebula,' because of its shape. It is a vast cloud of interstellar gas and is an easy object for binoculars or a small telescope.

(3) Albireo (α Cyg) is one of the most famous and most beautiful double stars in the sky. The components are strikingly different in color, one being a brilliant gold and the other a sapphire blue.

(4) The star 61 Cyg is famous as a very near star. At about 11 light years distant, it is the fourth nearest star visible to the unaided eye. It attained additional renown in 1838 when the German astronomer F. W. Bessel made it the first star whose trigonometric parallax was measured.

(5) Cygnus X-1 is an intense source of x-rays which many astronomers speculate could be a black hole. It is a very massive object (more than 10 solar masses) orbiting the 9th magnitude star HDE 226868.

(6) One of the most unusual nebulae in the sky is NGC 6826. In 1966, James Mullaney and Wallace McCall dubbed this object the 'Blinking Planetary.' With instruments of medium aperture, an observer looking directly at the 11th magnitude central star will not see the surrounding nebulosity. When averted vision is employed, however, the nebula becomes visible, masking the light from the star. Alternating between the two produces the 'blinking.'

Cygnus
Green, Jacob.
Astronomical
Recreations; or
Sketches of the
Relative Position
and Mythological
History of the
Constellations,
Philadelphia, 1824.

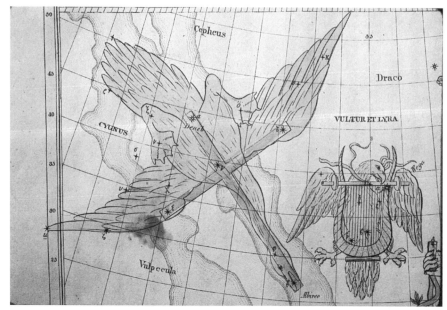

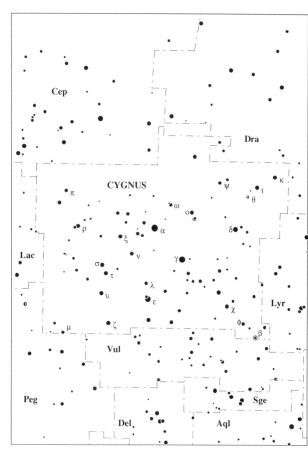

Key to Stellar magnitudes

- 5.6
- 5.3
- 5.0
- 4.5
- 3.8
- 3.1
- 2.8
- 2.5
- 2.0
- 1.0

197

Delphinus

Meaning:	The Dolphin (Porpoise)
Pronunciation:	dell fee' nus
Abbreviation:	Del
Possessive form:	Delphini (del fee' nee)
Asterisms:	Job's Coffin
Bordering constellations:	Aquarius, Aquila, Equuleus, Pegasus, Sagitta, Vulpecula
Overall brightness:	5.834 (61)
Central point:	RA = 20h39m Dec. = +11.5°
Directional extremes:	N = +21° S = +2° E = 21h06m W = 20h13m
Messier objects:	none
Meteor showers:	none
Midnight culmination date:	31 Jul
Bright stars:	none
Named stars:	Deneb (ε), Rotanev (β), Sualocin (α)
Near stars:	none
Size:	188.54 square degrees (0.457% of the sky)
Rank in size:	69
Solar conjunction date:	31 Jan
Visibility:	completely visible from latitudes: N of –69°
	completely invisible from latitudes: S of –88°
Visible stars:	(number of stars brighter than magnitude 5.5): 11
Interesting facts:	(1) The common names of α Del (Sualocin) and β Del (Rotanev) spelt backwards give the name of Nicolaus Venator, the assistant of the astronomer Giuseppe Piazzi.

Delphinus
Pardies, Ignace
Gaston. Globi
Coelestis in Tabulas
Planas Redacti
Descriptio, *Opus*
posthumus, Paris,
1674.

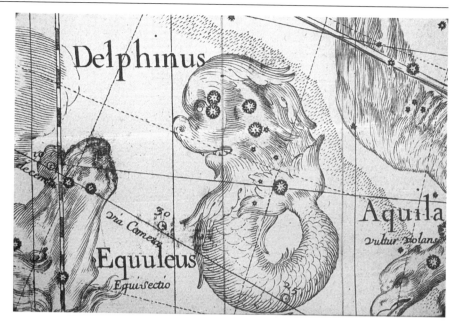

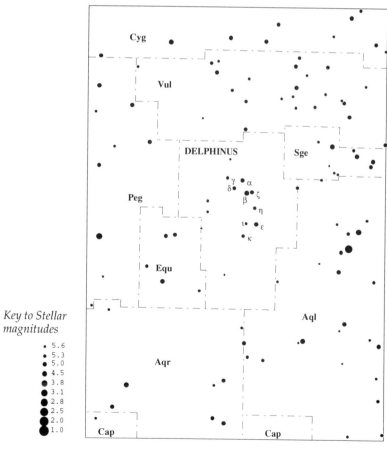

Key to Stellar
magnitudes

·	5.6
·	5.3
·	5.0
●	4.5
●	3.8
●	3.1
●	2.8
●	2.5
●	2.0
●	1.0

199

Dorado

Meaning:	The Swordfish
Pronunciation:	dor ah' doe
Abbreviation:	Dor
Possessive form:	Doradus (dor ah' dus)
Asterisms:	none
Bordering constellations:	Caelum, Horologium, Hydrus, Mensa, Pictor, Reticulum, Volans
Overall brightness:	8.372 (32)
Central point:	RA = 5h14m Dec. = –59.5°
Directional extremes:	N = –49° S = –70° E = 6h36m W = 3h52m
Messier objects:	none
Meteor showers:	none
Midnight culmination date:	17 Dec
Bright stars:	none
Named stars:	none
Near stars:	none
Size:	179.17 square degrees (0.434% of the sky)
Rank in size:	72
Solar conjunction date:	10 Jun
Visibility:	completely visible from latitudes: S of +20°
	completely invisible from latitudes: N of +41°
Visible stars:	(number of stars brighter than magnitude 5.5): 15

Interesting facts:

(1) This is one of 11 constellations invented by Pieter Dirksz Keyser and Frederick de Houtman, during the years 1595–7.

(2) One of the two satellite galaxies of the Milky Way lies mainly within the boundaries of this constellation, with some spillover into neighboring Mensa. This is the 'Large Magellanic Cloud' (LMC), also known by its Latin name 'Nebecula Major.' It was first recorded by a European in 1519, when Ferdinand Magellan noted it in his diary during his circumnavigation of the globe. The LMC is classified as an irregular galaxy, and it lies at an approximate distance of 180,000 light years.

(3) Within the LMC lies the largest diffuse nebula known, NGC 2070, commonly referred to as the 'Great Looped Nebula' or the 'Tarantula Nebula,' because of its resemblance to a spider in shape. To the unaided eye it appears as a faint star and was mistakenly given the designation 30 Doradus. If this nebula were placed at the distance of the more famous Orion Nebula (approximately 1500 light years) it would shine at magnitude –5.

(4) In February, 1987, a supernova became visible in the LMC. Dubbed 'Supernova 1987a,' this was the nearest such event since Kepler's Nova in 1604. It was also the brightest supernova by a factor of 1000 and the closest by a factor of 20 that had been studied in the telescopic age up to that point. It also proved to be a triumph for theoretical astrophysics because this was the first time neutrinos were detected from a supernova explosion.

(5) Also in Dorado, at RA = 6h Dec. = –66.5° lies the south ecliptic pole, near the border of the LMC.

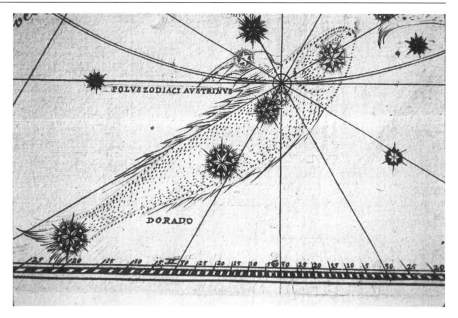

Dorado
Bayer, Johann.
Uranometria,
Augsburg, 1603. This
constellation was
invented by de
Houtman and Keyser
in 1596. It was first
illustrated on a globe
by Plancius, which has
not survived. This
photo from Bayer's
map, therefore, shows
the earliest existing
picture of this
constellation.

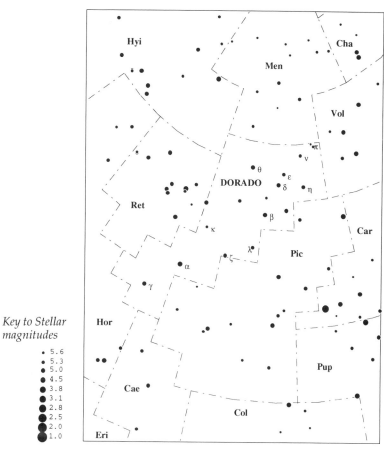

Key to Stellar
magnitudes

201

Draco

Meaning:	The Dragon
Pronunciation:	dray' koe
Abbreviation:	Dra
Possessive form:	Draconis (druh koe'niss)
Asterisms:	The Head, The Lozenge

Bordering constellations: Boötes, Camelopardalis, Cepheus, Cygnus, Hercules, Lyra, Ursa Major, Ursa Minor

Overall brightness:	7.295 (40)
Central point:	RA = 15h09m Dec. = +67°

Directional extremes: N = +86°   S = +48°   E = 21h00m   W = 9h18m

Messier objects:	none
Meteor showers:	Draconids (28 Jun)
	o Draconids (16 Jul)
	October Draconids (9 Oct)

Midnight culmination date: 24 May

Bright stars:	γ (66), η (116), β (125), δ (181)
Named stars:	Aldib (δ), Al Rakis (ν), Alsafi (σ), Altais (δ), Alwaid (β), Arrakis (μ), Dziban (ψ), Ed Asich (ι), Eltanin (γ), Etamin (γ), Giansar (λ), Grumium (ξ), Kuma (ν), Nodus I (ζ), Nodus II (δ), Rastaban (β), Thuban (α), Tyl (ε)
Near stars:	LFT 1431-1432 (16), BD+68°946 (31), σ Dra (49), x Dra (106), LFT 1552 (109), LTT 13665 (133)
Size:	1082.95 square degrees (2.625% of the sky)
Rank in size:	8

Solar conjunction date: 8 Nov

Visibility:	completely visible from latitudes: N of –4°
	completely invisible from latitudes: S of –42°
Visible stars:	(number of stars brighter than magnitude 5.5): 79

Interesting facts:
(1) α Dra, commonly known as Thuban, was the northern Pole Star approximately 5000 years ago. Its closest approach to the north celestial pole was probably around the year 2800 BC. Because of the effects of precession, α UMi (Polaris) now occupies that vaunted spot.

(2) The north ecliptic pole lies within this constellation. Its coordinates are RA = 18h Dec. = 66.5°.

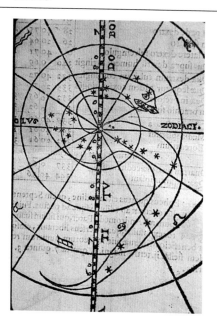

Draco
Gallucci, Giovanni
Paolo. Theatrum
Mundi, et Temporis,
Venice, 1588.

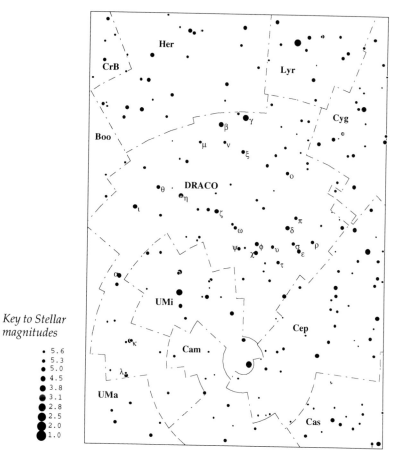

Key to Stellar
magnitudes

- 5.6
- 5.3
- 5.0
- 4.5
- 3.8
- 3.1
- 2.8
- 2.5
- 2.0
- 1.0

203

Equuleus

Meaning:	The Foal
Pronunciation:	ek woo oo' lee us
Abbreviation:	Equ
Possessive form:	Equulei (ek woo oo' lay ee)
Asterisms:	none
Bordering constellations:	Aquarius, Delphinus, Pegasus
Overall brightness:	6.979 (46)
Central point:	RA = 21h08m Dec. = +7.5°
Directional extremes:	N = +13° S = +2° E = 21h23m W = 20h54m
Messier objects:	none
Meteor showers:	none
Midnight culmination date:	8 Aug
Bright stars:	none
Named stars:	none
Near stars:	none
Size:	71.64 square degrees (0.174% of the sky)
Rank in size:	87
Solar conjunction date:	7 Feb
Visibility:	completely visible from latitudes: N of –77°
	completely invisible from latitudes: S of –88°
Visible stars:	(number of stars brighter than magnitude 5.5): 5

Interesting facts: (1) δ Equ is a close binary where both members are main sequence stars of spectral type F7, having visual magnitudes of 5.2 and 5.3. This pair appears to be separated by 0.34″. Since δ Equ lies approximately 53 light years away, it has been calculated that the distance between these two stars is only slightly more than the distance separating the Sun and Jupiter. These stars orbit one another every 5.7 years.

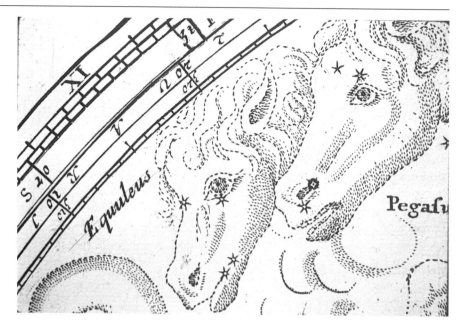

Equuleus
Sherburne, Edward.
The Sphere of
Marcus Manilius
made an English
poem: with
annotations and an
astronomical
appendix, *London,*
1675.

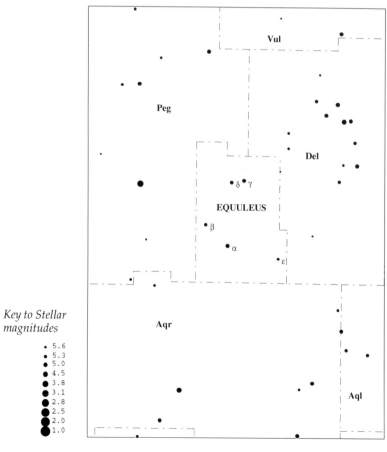

Key to Stellar
magnitudes

205

Eridanus

Meaning:	The River
Pronunciation:	air uh day' nus
Abbreviation:	Eri
Possessive form:	Eridani (air uh day' nee)
Asterisms:	none

Bordering constellations: Caelum, Cetus, Fornax, Horologium, Hydrus, Lepus, Orion, Phoenix, Taurus

Overall brightness:	6.942 (47)
Central point:	RA = 3h15m Dec. = –29°
Directional extremes:	N = 00° S = –58° E = 5h09m W = 1h22m
Messier objects:	none
Meteor showers:	none
Midnight culmination date:	10 Nov
Bright stars:	α (9), β (124), θ (151), γ (156)

Named stars: Acamar (θ), Achernar (α), Angetenar (τ^2), Azha (η), Beid (o^1), Cursa (β), Keid (o^2), Rana (δ), Sceptrum (53), Theemin (υ^2), Zaurak (γ), Zibel (ζ)

Near stars: ε Eri (10), 40 Eri A-B-C (38), 82 Eri (69), ρ Eri A-B (77), BD–13°544 (111), δ Eri (148), BD–5° 1123 (156) LTT 1830–1831 (184)

Size:	1137.92 square degrees (2.758% of the sky)
Rank in size:	6
Solar conjunction date:	11 May
Visibility:	completely visible from latitudes: S of +32° portions visible worldwide
Visible stars:	(number of stars brighter than magnitude 5.5): 79

Interesting facts: (1) ε Eri is a very nearby solar-type star, lying at a distance of 10.7 light years. For readers of the 'Star Trek' science fiction series, this is the star around which the planet Vulcan (Mr. Spock's home planet) supposedly revolves.

Eridanus
Seller, John. Atlas
Coelestis..., *London,*
1700.

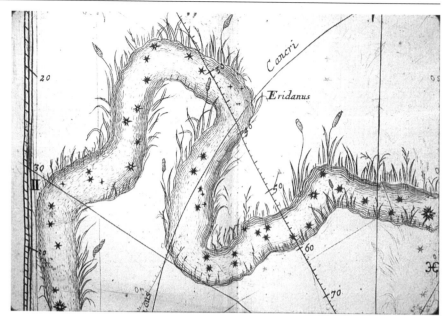

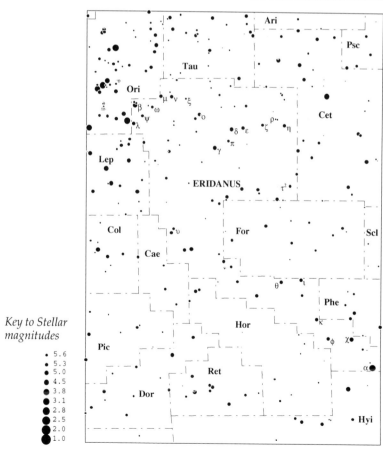

Key to Stellar
magnitudes

Fornax

Meaning: The Laboratory Furnace

Pronunciation: for' nax

Abbreviation: For

Possessive form: Fornacis (for nay' siss)

Asterisms: none

Bordering constellations: Cetus, Eridanus, Phoenix, Sculptor

Overall brightness: 3.019 (87)

Central point: RA = 2h46m   Dec. = −32°

Directional extremes: N = −24°   S = −40°   E = 3h48m   W = 1h44m

Messier objects: none

Meteor showers: none

Midnight culmination date: 2 Nov

Bright stars: none

Named stars: none

Near stars: LFT 193 (147)

Size: 397.50 square degrees (0.964% of the sky)

Rank in size: 41

Solar conjunction date: 4 May

Visibility: completely visible from latitudes: S of +50°
completely invisible from latitudes: N of +64°

Visible stars: (number of stars brighter than magnitude 5.5):   12

Interesting facts: (1) This is one of 14 constellations which Lacaille devised while at the Cape of Good Hope in 1751–2.

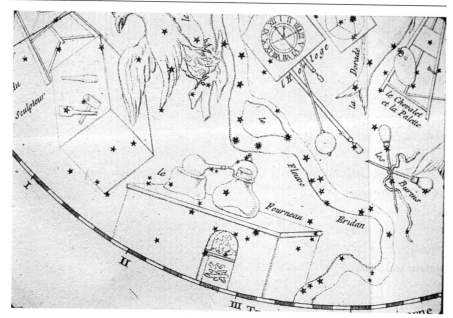

Fornax (labeled 'le Fourneau' on this map)
Lacaille, Nicolas Louis de. Planisphere contenant les Constellations Celestes, found in Mémoires Académie Royale des Sciences, *Paris, 1752 (published in 1756). This constellation was invented by Lacaille and the photo shows its first appearance on any star map.*

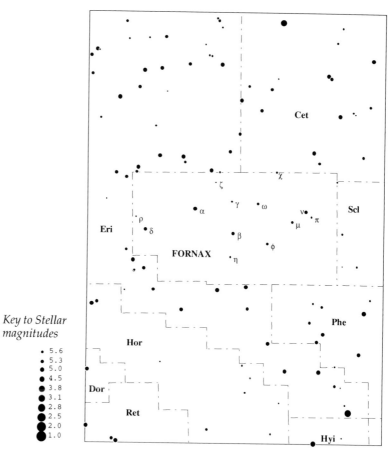

Key to Stellar magnitudes

Gemini

Meaning:	The Twins
Pronunciation:	gem' in eye
Abbreviation:	Gem
Possessive form:	Geminorum (jem uh nor' um)
Asterisms:	The Heavenly G, The Winter Octagon, The Winter Oval

Bordering constellations: Auriga, Cancer, Canis Minor, Lynx, Monoceros, Orion, Taurus

Overall brightness:	9.148 (26)
Central point:	RA = 7h01m Dec. = +22.5°
Directional extremes:	N = +35° S = +10° E = 8h06m W = 5h57m
Messier objects:	M35
Meteor showers:	ε Geminids (19 Oct)
	Geminids (14 Dec)

Midnight culmination date: 5 Jan

Bright stars:	β (17), α (23), γ (43), μ (143), ε (161)
Named stars:	Alhena (γ), Almeisan (γ), Castor (α), Mebsuta (ε), Mekbuda (ζ), Pollux (β), Propus (η), Tejat Posterior (μ), Tejat Prior (η)
Near stars:	Wolf 294 (63), Ross 64 (130), Wolf 287 (170)
Size:	513.76 square degrees (1.245% of the sky)
Rank in size:	30

Solar conjunction date: 8 Jul

Visibility:	completely visible from latitudes: N of –55°
	completely invisible from latitudes: S of –80°
Visible stars:	(number of stars brighter than magnitude 5.5): 47

Interesting facts: (1) Two of the three planet 'discoveries' have occurred within the boundaries of this constellation. In 1781, William Herschel spotted the planet Uranus near η Gem. Clyde Tombaugh, working at the Lowell Observatory in Flagstaff, Arizona, 149 years later exposed a series of plates centered on the star δ Gem and found the planet Pluto.

(2) α Gem, also known as Castor, is a complex system of six stars which appear as one to the unaided eye. Although officially designated as a triple star, further analysis has shown that each of the three components is itself a double star. This is one of the most complex systems in the sky, and studies indicate that systems of more than six stars would quickly become unstable and separate.

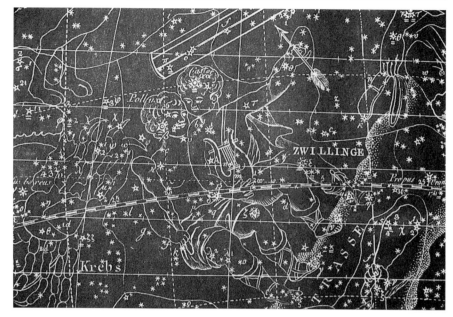

Gemini (labeled 'Zwillinge' on this map)
Goldbach, Christian Friedrich. Neuester Himmels – Atlas zum Gebrauche für Schul – und Akademischen Unterricht, nach Flamsteed, *Weiman, 1799.*

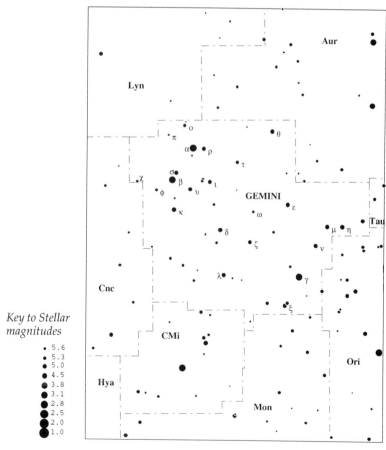

Key to Stellar magnitudes

211

Grus

Meaning:	The Crane
Pronunciation:	groose
Abbreviation:	Gru
Possessive form:	Gruis (groo' eese)
Asterisms:	none

Bordering constellations: Indus, Microscopium, Phoenix, Piscis Austrinus, Sculptor, Tucana

Overall brightness:	6.566 (50)
Central point:	RA = 22h25m Dec. = −47°

Directional extremes: N = −37°   S = −57°   E = 23h25m   W = 21h25m

Messier objects:	none
Meteor showers:	none

Midnight culmination date: 28 Aug

Bright stars:	α (30), β (57), γ (172)
Named stars:	Al Dhanab (γ), Alnair (α)
Near stars:	LFT 1640 (32)
Size:	365.51 square degrees (0.886% of the sky)
Rank in size:	45

Solar conjunction date: 27 Feb

Visibility:	completely visible from latitudes: S of +33°
	completely invisible from latitudes: N of +53°
Visible stars:	(number of stars brighter than magnitude 5.5): 24

Interesting facts: (1) This is one of 11 constellations invented by Pieter Dirksz Keyser and Frederick de Houtman, during the years 1595–7.

Grus
Bayer, Johann.
Uranometria,
Augsburg, 1603. This
constellation was
invented by de
Houtman and Keyser
in 1596. It was first
illustrated on a globe
by Plancius, which has
not survived. This
photo from Bayer's
map, therefore, shows
the earliest existing
picture of this
constellation.

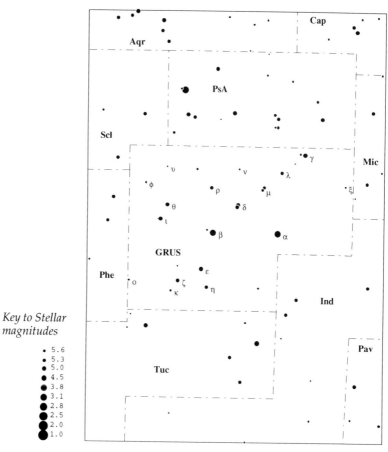

Key to Stellar
magnitudes

Hercules

Meaning:	Hercules (the hero)
Pronunciation:	her' cue leez
Abbreviation:	Her
Possessive form:	Herculis (her' kyoo liss)
Asterisms:	The Butterfly, The Keystone

Bordering constellations: Aquila, Boötes, Corona Borealis, Draco, Lyra, Ophiuchus, Sagitta, Serpens, Vulpecula

Overall brightness:	6.938 (48)
Central point:	RA = 17h21m Dec. = +27.5°
Directional extremes:	N = +51° S = +4° E = 18h56m W = 15h47m
Messier objects:	M13, M92
Meteor showers:	τ Herculids (3 Jun)

Midnight culmination date: 13 Jun

Bright stars:	ζ (129), α¹ (183), δ (197), π (198)
Named stars:	Cujam (ω), Kajam (ι), Kornephoros (β), Maasym (λ), Marfak (κ), Marsik (κ), Ras Algethi (α), Sarin (δ)
Near stars:	LFT 1326-1327 (75), LFT 1273 (92), μ Her A-B-C (107), LFT 1363 (113), ζ Her A-B (174), BD+33°2777 (175), LFT 1371 (182), Ross 863 (187)
Size:	1225.15 square degrees (2.970% of the sky)
Rank in size:	5

Solar conjunction date: 12 Dec

Visibility:	completely visible from latitudes: N of −39°
	completely invisible from latitudes: S of −86°
Visible stars:	(number of stars brighter than magnitude 5.5): 85

Interesting facts: (1) The solar apex, that point on the celestial sphere toward which the sun seems to be moving (due to its motion within the Milky Way) is found within this constellation. The approximate coordinates of the solar apex are RA = 18h Dec. = +34°, about 3° south of the star θ Her.

(2) The Great Globular Cluster in Hercules is one of the finest objects of its kind in the sky, surpassed only by the globular clusters ω Cen and 47 Tuc, both of which are located in the far southern sky. This cluster, also known as M13 and NGC 6205, was discovered in 1714 by Edmund Halley, who also noted that it was visible to the unaided eye in a dark sky. Its magnitude is 5.7 and it can be found by looking ⅓ of the way from η Her to ζ Her. It lies at an approximate distance of 25 000 light years, and although estimates of the number of stars it contains vary, it can safely be said that this object is composed of more than 100 000 stars.

(3) α Her is a wonderful example of a colorful double star. The primary of this pair is orange, and the fainter secondary star is – to this writer – olive green, although some very fine observers report seeing brighter shades of green in the light of this star.

Hercules
Burritt, Elijah H.
Atlas Designed to
Illustrate the
Geography of the
Heavens, *Hartford,*
1835.

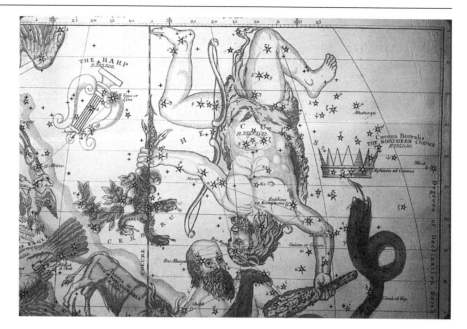

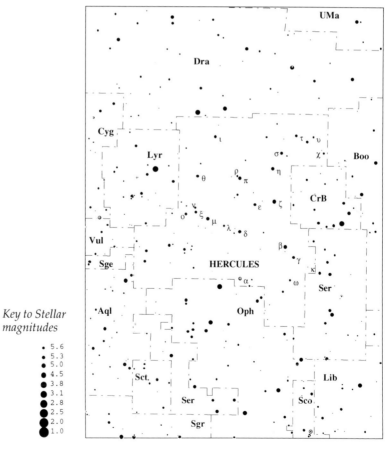

Key to Stellar
magnitudes

- 5.6
- 5.3
- 5.0
- 4.5
- 3.8
- 3.1
- 2.8
- 2.5
- 2.0
- 1.0

215

Horologium

Meaning:	The Pendulum Clock
Pronunciation:	hor uh low' gee um
Abbreviation:	Hor
Possessive form:	Horologii (hor owe low' gee ee)
Asterisms:	none

Bordering constellations: Caelum, Dorado, Eridanus, Hydrus, Reticulum

Overall brightness:	4.018 (81)
Central point:	RA = 3h15m Dec. = −53.5°
Directional extremes:	N = −40° S = −67° E = 4h18m W = 2h12m
Messier objects:	none
Meteor showers:	none

Midnight culmination date: 10 Nov

Bright stars:	none
Named stars:	none
Near stars:	none
Size:	248.88 square degrees (0.603% of the sky)
Rank in size:	58

Solar conjunction date: 11 May

Visibility:	completely visible from latitudes: S of +23°
	completely invisible from latitudes: N of +50°
Visible stars:	(number of stars brighter than magnitude 5.5): 10
Interesting facts:	(1) This was one of the 14 constellations invented by Lacaille during his stay at the Cape of Good Hope in 1751–2.

Horologium (labeled 'l'Horloge' on this map)
Lacaille, Nicolas Louis de. Planisphere contenant les Constellations Celestes, found in Mémoires Académie Royale des Sciences, *Paris, 1752 (published in 1756). This constellation was invented by Lacaille and the photo shows its first appearance on any star map.*

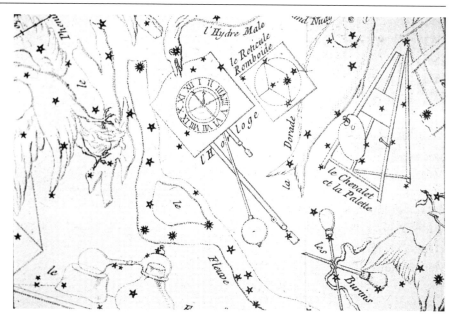

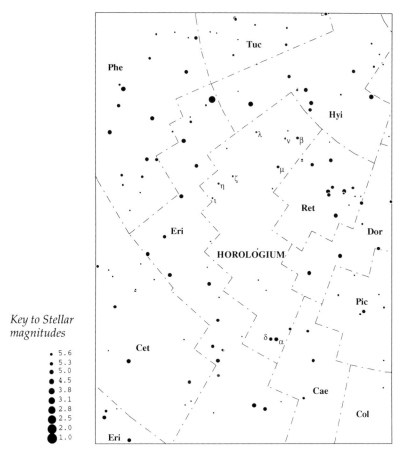

Key to Stellar magnitudes

- 5.6
- 5.3
- 5.0
- 4.5
- 3.8
- 3.1
- 2.8
- 2.5
- 2.0
- 1.0

217

Hydra

Meaning:	The Water Snake
Pronunciation:	hi' druh
Abbreviation:	Hya
Possessive form:	Hydrae (hide' rye)
Asterisms:	The Head

Bordering constellations: Antlia, Cancer, Canis Minor, Centaurus, Corvus, Crater, Leo, Libra, Monoceros, Puppis, Pyxis, Sextans, Virgo

Overall brightness:	5.449 (71)
Central point:	RA = 11h33m Dec. = −14°
Directional extremes:	N = +7° S = −35° E = 14h58m W = 8h08m
Messier objects:	M48, M68, M83
Meteor showers:	σ Hydrids (11 Dec)
Midnight culmination date:	15 Mar
Bright stars:	α (46), γ (169), ζ (186), ν (187)
Named stars:	Alphard (α)
Near stars:	LFT 661 (117), BD-12°2918 A-B (151), LFT 823 A-B (166), LFT 865 (185)
Size:	1302.84 square degrees (3.158% of the sky)
Rank in size:	1
Solar conjunction date:	15 Sep
Visibility:	completely visible from latitudes: +55° to −83°
	portions visible worldwide
Visible stars:	(number of stars brighter than magnitude 5.5): 71

Interesting facts: (1) V Hya, a variable star within this constellation, is often considered the reddest known star. At a maximum visual magnitude of 6.5, at least a small telescope is required to observe this unusual object. Another variable, U Hya, is somewhat brighter (4.7–6.2) and nearly as red. Both stars are within 5° of ν Hya, with V Hya lying to the south and U Hya to the northwest.

(2) On 18 September 1965, one of the most famous comets of the twentieth century was discovered near α Hya. This was the sungrazing comet Ikeya–Seki, which, on 21 October 1965, was visible in daylight when only 2° from the Sun.

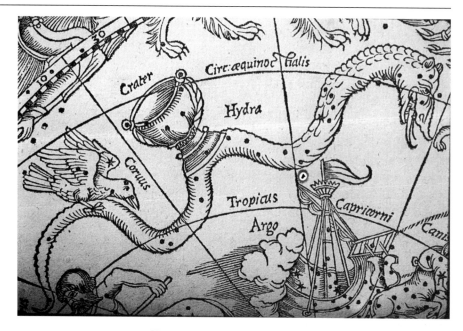

Hydra
Ptolemy. Omnia,
quae extant, opera,
Basel, 1541.

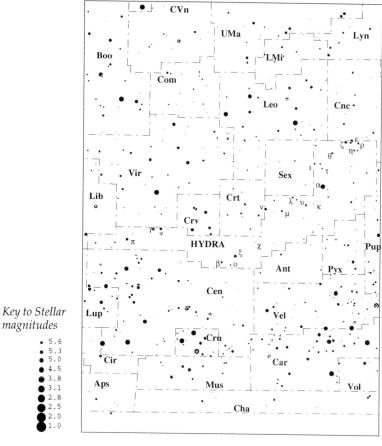

*Key to Stellar
magnitudes*

- 5.6
- 5.3
- 5.0
- 4.5
- 3.8
- 3.1
- 2.8
- 2.5
- 2.0
- 1.0

Hydrus

Meaning:	The Southern Water Snake
Pronunciation:	hi' druss
Abbreviation:	Hyi
Possessive form:	Hydri (hide' ree)
Asterisms:	none

Bordering constellations: Dorado, Eridanus, Horologium, Mensa, Octans, Reticulum, Tucana

Overall brightness:	5.760 (63)
Central point:	RA = 2h16m Dec. = −70°

Directional extremes: N = −58°   S = −82°   E = 4h33m   W = 00h02m

Messier objects:	none
Meteor showers:	none

Midnight culmination date: 26 Oct

Bright stars:	β (126), α (138)
Named stars:	none
Near stars:	β Hyi (71)
Size:	243.04 square degrees (0.589% of the sky)
Rank in size:	61

Solar conjunction date: 26 Apr

Visibility:	completely visible from latitudes: S of +8° completely invisible from latitudes: N of +32°
Visible stars:	(number of stars brighter than magnitude 5.5): 14

Interesting facts: (1) This is one of 11 constellations invented by Pieter Dirksz Keyser and Frederick de Houtman, during the years 1595–7.

(2) Hydrus is one of only two constellations whose abbreviation (Hyi) contains a letter ('i') not found in the constellation name. The other is Sagitta (Sge).

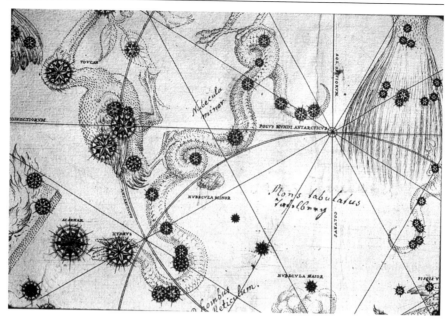

*Hydrus
Bayer, Johann.*
Uranometria,
*Augsburg, 1603. This
constellation was
invented by de
Houtman and Keyser
in 1596. It was first
illustrated on a globe
by Plancius, which has
not survived. This
photo from Bayer's
map, therefore, shows
the earliest existing
picture of this
constellation.*

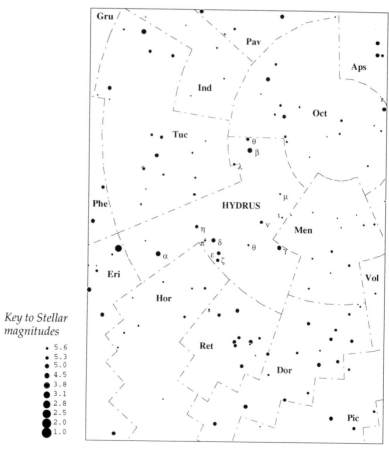

*Key to Stellar
magnitudes*

- 5.6
- 5.3
- 5.0
- 4.5
- 3.8
- 3.1
- 2.8
- 2.5
- 2.0
- 1.0

Indus

Meaning:	The American Indian
Pronunciation:	in′ dus
Abbreviation:	Ind
Possessive form:	Indi (in′ dee)
Asterisms:	none
Bordering constellations:	Grus, Microscopium, Octans, Pavo, Telescopium, Tucana
Overall brightness:	4.422 (80)
Central point:	RA = 21h55m Dec. = −60°
Directional extremes:	N = −45° S = −75° E = 23h25m W = 20h25m
Messier objects:	none
Meteor showers:	none
Midnight culmination date:	12 Aug
Bright stars:	α (189)
Named stars:	none
Near stars:	ε Ind (14)
Size:	294.01 square degrees (0.713% of the sky)
Rank in size:	49
Solar conjunction date:	19 Feb
Visibility:	completely visible from latitudes: S of +15° completely invisible from latitudes: N of +45°
Visible stars:	(number of stars brighter than magnitude 5.5): 13
Interesting facts:	(1) This is one of 11 constellations invented by Pieter Dirksz Keyser and Frederick de Houtman, during the years 1595–7. (2) One of the nearest of the solar-type stars ε Ind, lies within the confines of this constellation. Its distance is 11.2 light years.

Indus
Bayer, Johann.
Uranometria,
Augsburg, 1603. This
constellation was
invented by de
Houtman and Keyser
in 1596. It was first
illustrated on a globe
by Plancius, which has
not survived. This
photo from Bayer's
map, therefore, shows
the earliest existing
picture of this
constellation.

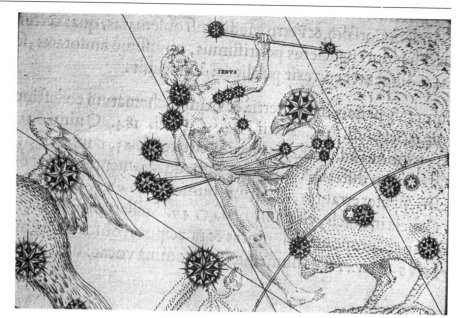

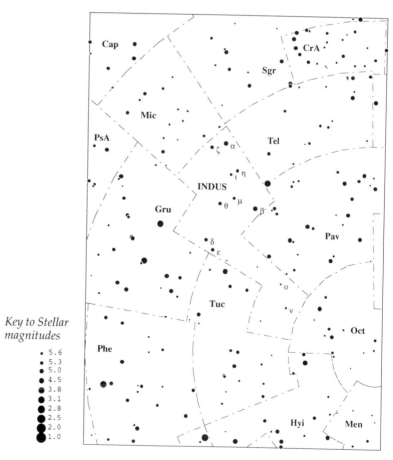

Key to Stellar
magnitudes

. 5.6
. 5.3
. 5.0
. 4.5
. 3.8
. 3.1
. 2.8
. 2.5
. 2.0
. 1.0

223

Lacerta

Meaning:	The Lizard
Pronunciation:	luh sir' tuh
Abbreviation:	Lac
Possessive form:	Lacertae (luh sir' tie)
Asterisms:	none

Bordering constellations: Andromeda, Cassiopeia, Cepheus, Cygnus, Pegasus

Overall brightness:	11.460 (13)
Central point:	RA = 22h25m Dec. = +46°

Directional extremes: N = +57°   S = +35°   E = 22h56m   W = 21h55m

Messier objects:	none
Meteor showers:	none

Midnight culmination date: 28 Aug

Bright stars:	none
Named stars:	none
Near stars:	EV Lac (43)
Size:	200.69 square (0.487%)
Rank in size:	68

Solar conjunction date: 27 Feb

Visibility:	completely visible from latitudes: N of –33°
	completely invisible from latitudes: S of –55°
Visible stars:	(number of stars brighter than magnitude 5.5): 23

Interesting facts: (1) One of seven constellations still in use invented by Johannes Hevelius. In 1690, this group was included in a star atlas which accompanied his stellar catalog.

(2) A very unusual object lies within this constellation. It is designated BL Lac, because it was originally believed to be a variable star. It is now known that this object, and many others like it, are very distant, violently variable objects which resemble quasars in size and energy output. BL Lac type objects differ from quasars, however, because they seem to be related to distant elliptical galaxies and because their spectra show no discernable lines.

Lacerta
Hevelius, Johannes.
Firmamentum
Sobiescianum, sive
Uranographia,
totum Coelum
Stellatum, *Gdansk,*
1690. This is the first
appearance of this
constellation on any
star map.

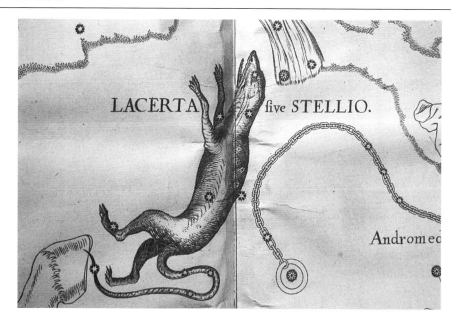

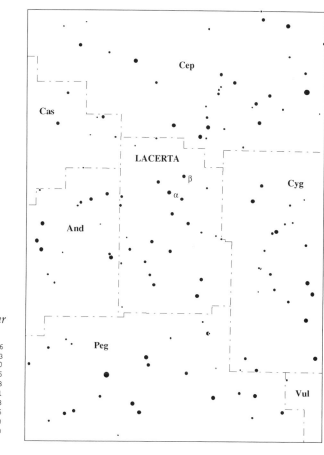

Key to Stellar
magnitudes

5.6
5.3
5.0
4.5
3.8
3.1
2.8
2.5
2.0
1.0

Leo

Meaning:	The Lion
Pronunciation:	lee' owe
Abbreviation:	Leo
Possessive form:	Leonis (lee owe' niss)
Asterisms:	The Diamond (of Virgo), The Sickle, The Spring Triangle

Bordering constellations: Cancer, Coma Berenices, Crater, Hydra, Leo Minor, Sextans, Ursa Major, Virgo

Overall brightness:	5.491 (70)
Central point:	RA = 10h37m Dec. = +13.5°
Directional extremes:	N = +33° S = –6° E = 11h56m W = 9h18m
Messier objects:	M65, M66, M95, M96, M105
Meteor showers:	δ Leonids (26 Feb)
	σ Leonids (17 Apr)
	Leonids (17 Nov)

Midnight culmination date: 1 Mar

Bright stars:	α (21), γ (41), β (59), δ (94), ε (162)
Named stars:	Adhafera (ζ), Algieba (γ), Alterf (λ), Chort (θ), Coxa (θ), Denebola (β), Ras Elased Australis (ε), Ras Elased Boraelis (μ), Regulus (α), Subra (ο), Zosma (δ)
Near stars:	Wolf 359 (4), AD Leo (37), Ross 104 (79), Wolf 358 (87), Ross 905 (144)
Size:	946.96 square degrees (2.296% of the sky)
Rank in size:	12

Solar conjunction date: 31 Aug

Visibility:	completely visible from latitudes: +84° to –57°
	portions visible worldwide
Visible stars:	(number of stars brighter than magnitude 5.5): 52

Non-traditional 'mythology': Although the figure of a lion is easily seen in the stars of Leo, often it is reported that a mouse (or rat) may be glimpsed. For the tip of the nose, use β Leo; for the pointed head connect β Leo to both δ Leo and θ Leo. A line from θ Leo to α Leo to γ Leo and back to δ Leo traces out the body, and the curly tail is the top of the Sickle (ζ, μ, and ε).

In addition, rather than the asterism of the Sickle, many observers correctly note that these stars represent a backward question mark.

Interesting facts: (1) α Leo, or Regulus, is one of the four Royal Stars of the ancient Persians. The other three are Aldebaran (α Tau), Antares (α Sco), and Fomalhaut (α PsA).

(2) The Leonid meteor shower, which peaks each year around 17 Nov, is unusually active every 33 years. Tremendous displays were noted in 1799, 1833, 1866, and as recently as 1966. Tens of thousands of meteors per hour have been recorded on these occasions. The next great shower is expected in 1999, and as the Moon will have just passed its first quarter phase the night before, the view should be unhampered in the early morning hours when the shower is expected to peak.

Leo
Wollaston, Francis. A Portrature of the Heavens, as They Appear to the Naked Eye, *London, 1811.*

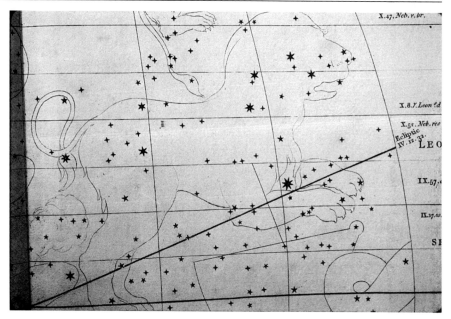

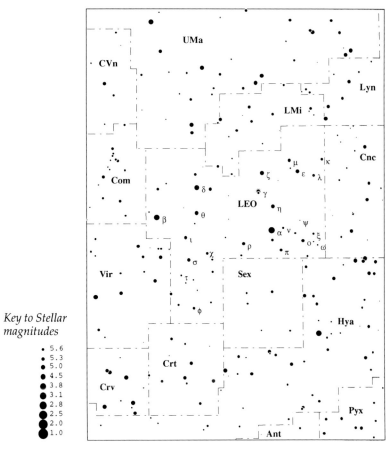

Key to Stellar magnitudes

- 5.6
- 5.3
- 5.0
- 4.5
- 3.8
- 3.1
- 2.8
- 2.5
- 2.0
- 1.0

227

Leo Minor

Meaning:	The Lion Cub
Pronunciation:	lee' owe my' nor
Abbreviation:	LMi
Possessive form:	Leonis Minoris (lee owe' niss my nor' iss)
Asterisms:	none

Bordering constellations: Leo, Lynx, Ursa Major

Overall brightness:	6.467 (52)
Central point:	RA = 10h11m Dec. = +32.5°
Directional extremes:	N = +42° S = +23° E = 11h04m W = 9h19m
Messier objects:	none
Meteor showers:	Leo Minorids (24 Oct)

Midnight culmination date: 23 Feb

Bright stars:	none
Named stars:	Praecipua (46)
Near stars:	11 LMi A-B (157)
Size:	231.96 square degrees (0.562% of the sky)
Rank in size:	64

Solar conjunction date: 25 Aug

Visibility:	completely visible from latitudes: N of –48°
	completely invisible from latitudes: S of –67°
Visible stars:	(number of stars brighter than magnitude 5.5): 15
Interesting facts:	(1) One of seven constellations still in use invented by Johannes Hevelius. In 1690, this group was included in a star atlas which accompanied his stellar catalog.

Leo Minor
Hevelius, Johannes.
*Firmamentum
Sobiescianum, sive
Uranographia,
totum Coelum
Stellatum, Gdansk,
1690. This is the first
appearance of this
constellation on any
star map.*

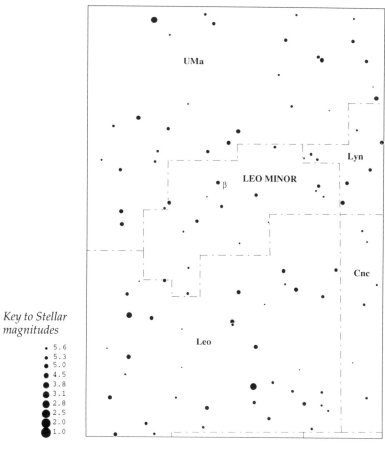

*Key to Stellar
magnitudes*

. 5.6
. 5.3
. 5.0
. 4.5
● 3.8
● 3.1
● 2.8
● 2.5
● 2.0
● 1.0

Lepus

Meaning:	The Hare
Pronunciation:	lee' pus
Abbreviation:	Lep
Possessive form:	Leporis (lee por' iss)
Asterisms:	none

Bordering constellations: Caelum, Canis Major, Columba, Eridanus, Monoceros, Orion

Overall brightness: 9.646 (23)

Central point: RA = 5h31m   Dec. = –19°

Directional extremes: N = –11°   S = –27°   E = 6h09m   W = 4h54m

Messier objects: M79

Meteor showers: none

Midnight culmination date: 14 Dec

Bright stars: α (96), β (134)

Named stars: Arneb (α), Nihal (β)

Near stars: BD–21°1377 (53), BD–21°1051 A-B (100), γ Lep A-B-C (116)

Size: 290.29 square degrees (0.704% of the sky)

Rank in size: 51

Solar conjunction date: 15 Jun

Visibility: completely visible from latitudes: S of +63°
completely invisible from latitudes: N of +79°

Visible stars: (number of stars brighter than magnitude 5.5): 28

Interesting facts: (1) Approximately 5° east and 2° north of μ Lep is the star R Lep, also known as 'Hind's Crimson Star.' The astronomer J. R. Hind noted its intense red color in 1845, comparing it to a drop of blood against a black sky. The magnitude range of this star varies from 6th to about 10th.

Lepus
Bevis, John.
Uranographia
Britannica, *London,*
1745.

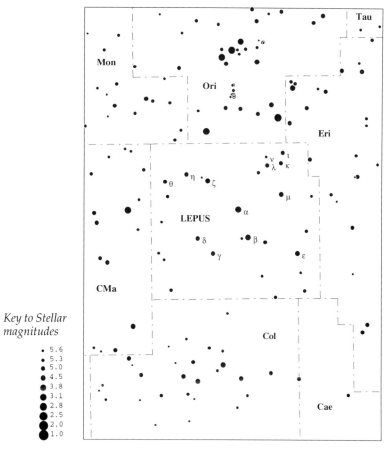

*Key to Stellar
magnitudes*

· 5.6
· 5.3
· 5.0
• 4.5
● 3.8
● 3.1
● 2.8
● 2.5
● 2.0
● 1.0

231

Libra

Meaning:	The Scales
Pronunciation:	lye' bruh
Abbreviation:	Lib
Possessive form:	Librae (lye' bry)
Asterisms:	none

Bordering constellations: Hydra, Lupus, Ophiuchus, Scorpius, Serpens, Virgo

Overall brightness:	6.505 (51)
Central point:	RA = 15h08m Dec. = –15°

Directional extremes: N = 00°   S = –30°   E = 15h59m   W = 14h18

Messier objects:	none
Meteor showers:	Librids (8 Jun)

Midnight culmination date: 9 May

Bright stars:	β (100), α² (119)
Named stars:	Kiffa Australis (α), Kiffa Boraelis (β), Zubenelakrab (γ), Zubenelakribi (δ), Zubenelgenubi (α), Zubeneschamali (β), Zuben Hakrabi (ν)
Near stars:	ADS 9446 A-B (46), BD–11°3759 (70), Wolf 562 (76), LFT 1218 (150)
Size:	538.05 square degrees (1.304% of the sky)
Rank in size:	29

Solar conjunction date: 8 Nov

Visibility:	completely visible from latitudes: S of +60°
	portions visible worldwide
Visible stars:	(number of stars brighter than magnitude 5.5): 35
Interesting facts: (1)	β Lib is the only star visible to the unaided eye which has a decidedly green tint. This is disputed by some observers, but many years of observing this star and questioning individuals with both trained and untrained eyes has convinced this writer of the validity of the above statement.

Libra
Bode, Johann Elert.
Uranographia Sive
Astrorum
Descriptio, *Berlin,*
1801.

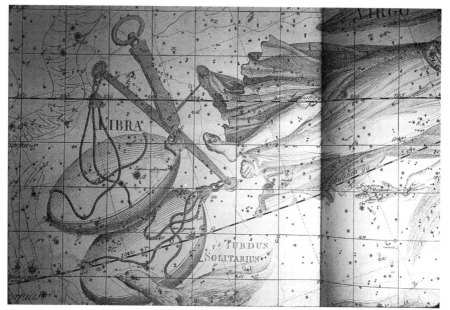

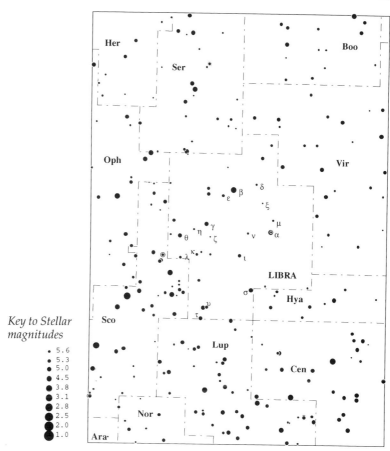

Key to Stellar
magnitudes

Lupus

Meaning:	The Wolf
Pronunciation:	loo′ pus
Abbreviation:	Lup
Possessive form:	Lupi (loo′ pee)
Asterisms:	none

Bordering constellations: Centaurus, Circinus, Libra, Norma, Scorpius

Overall brightness:	14.984 (5)
Central point:	RA = 15h09m Dec. = –42.5°
Directional extremes:	N = –30° S = –55° E = 16h05m W = 14h13m
Messier objects:	none
Meteor showers:	none

Midnight culmination date: 9 May

Bright stars:	α (74), β (107), γ (123)
Named stars:	none
Near stars:	LFT 1208 (61)
Size:	333.68 square degrees (0.809% of the sky)
Rank in size:	46

Solar conjunction date: 8 Nov

Visibility:	completely visible from latitudes: S of +35°
	completely invisible from latitudes: N of +60°
Visible stars:	(number of stars brighter than magnitude 5.5): 50

Interesting facts: (1) One of the brightest supernova explosions which has occurred in our galaxy was seen near β Lup in the year 1006. Historical accounts estimate the brightness as 'three times as bright as Venus,' and 'a quarter the brightness of the Moon.' These and other indications place the visual magnitude at approximately –8 to –10. This is the only supernova to be recorded in Europe and the Arab empire before the Renaissance.

Lupus (labeled 'Fera' on this map) Aratus Solensis. Phainomena kai Diosemeia, *Oxford, 1672.*

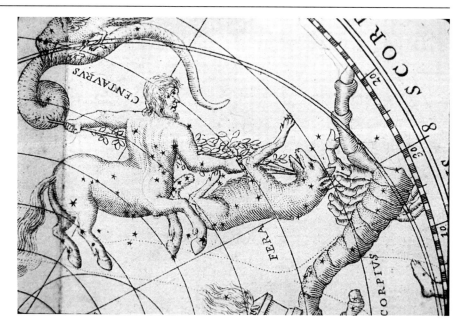

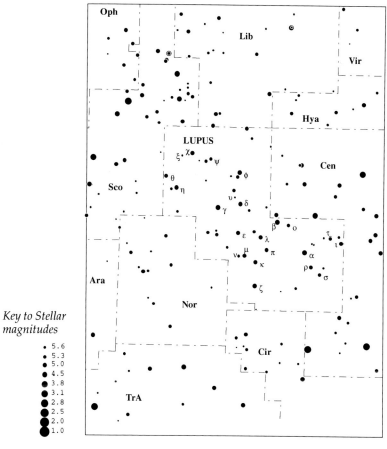

Key to Stellar magnitudes

Lynx

Meaning:	The Lynx
Pronunciation:	links
Abbreviation:	Lyn
Possessive form:	Lyncis (lin' siss)
Asterisms:	none

Bordering constellations: Auriga, Camelopardalis, Cancer, Gemini, Leo Minor, Ursa Major

Overall brightness:	5.684 (66)
Central point:	RA = 7h56m Dec. = +46.5°
Directional extremes:	N = +62° S = +33° E = 9h40m W = 6h13m
Messier objects:	none
Meteor showers:	none
Midnight culmination date:	19 Jan
Bright stars:	α (191)
Named stars:	none
Near stars:	none
Size:	545.39 square degrees (1.322% of the sky)
Rank in size:	28
Solar conjunction date:	22 Jul
Visibility:	completely visible from latitudes: N of –28°
	completely invisible from latitudes: S of –57°
Visible stars:	(number of stars brighter than magnitude 5.5): 31

Interesting facts:
(1) One of seven constellations still in use invented by Johannes Hevelius. In 1690, this group was included in a star atlas which accompanied his stellar catalog.

(2) The star 41 Lyn, numbered by Flamsteed in the early eighteenth century, has since moved into the constellation of Ursa Major. Although this is far from a unique occurrence, this is the star usually singled out to demonstrate stellar motion and to prove the fragility of constellation boundaries.

(3) Exactly 7° north of Castor (α Gem) is the most distant globular cluster in our galaxy, NGC 2419. Sometimes called the 'Intergalactic Wanderer,' this cluster's distance of 182 000 light years rivals that of the Magellanic Clouds.

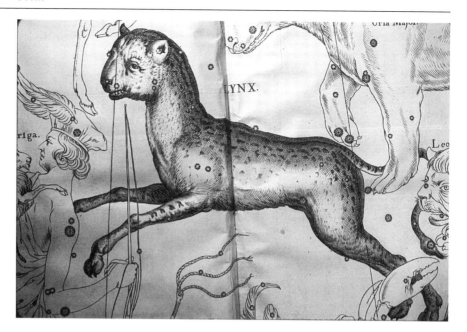

Lynx
Hevelius, Johannes.
Firmamentum
Sobiescianum, sive
Uranographia,
totum Coelum
Stellatum, *Gdansk,*
1690. This is the first
appearance of this
constellation on any
star map.

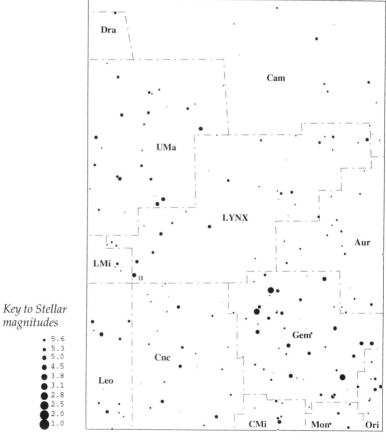

Key to Stellar
magnitudes

- 5.6
- 5.3
- 5.0
- 4.5
- 3.8
- 3.1
- 2.8
- 2.5
- 2.0
- 1.0

Lyra

Meaning:	The Harp
Pronunciation:	lie' ruh
Abbreviation:	Lyr
Possessive form:	Lyrae (lie' rye)
Asterisms:	The Summer Triangle

Bordering constellations: Cygnus, Draco, Hercules, Vulpecula

Overall brightness:	9.076 (28)
Central point:	RA = 18h49m Dec. = +36.5°

Directional extremes: N = +48°   S = +25°   E = 19h26m   W = 18h12m

Messier objects:	M56, M57
Meteor showers:	April Lyrids (22 Apr)
	June Lyrids (16 Jun)

Midnight culmination date: 4 Jul

Bright stars:	α (5)
Named stars:	Aladfar (η), Sheliak (β), Sulaphat (γ), Vega (α)
Near stars:	α Lyr (115), 17 Lyr C (118)
Size:	286.48 square degrees (0.694% of the sky)
Rank in size:	52

Solar conjunction date: 3 Jan

Visibility:	completely visible from latitudes: N of –42°
	completely invisible from latitudes: S of –65°
Visible stars:	(number of stars brighter than magnitude 5.5): 26

Interesting facts:
(1) Vega (α) was the first star to be photographed (1850) and also the first to have its spectrum photographed (1872). It was also one of the first three stars to have its parallax measured (probably the third). This was accomplished by Struve in 1840.
(2) Vega is the only single star in the sky to have an automobile named for it, the Chevrolet Vega.
(3) Since the mid-nineteenth century, Vega has been used as the standard zero-magnitude star. The actual visual magnitude of this star is extremely close to zero: 0.03. It is also the standard spectral type A0 main sequence star.
(4) Because of precession, Vega will be the brightest star near the north celestial pole in approximately 14 000 AD.
(5) The IRAS satellite (1983) detected excess infrared radiation around Vega. This seems to be an embryonic planetary system in the early stages of development.
(6) ε Lyr is the so-called 'Double-Double' star. The fact that four stars occupy this space was first detected in 1779 by William Herschel. An observer with good eyes can separate ε into two distinct 'stars,' but a telescope of 60 mm employing medium power is necessary to see all four components.

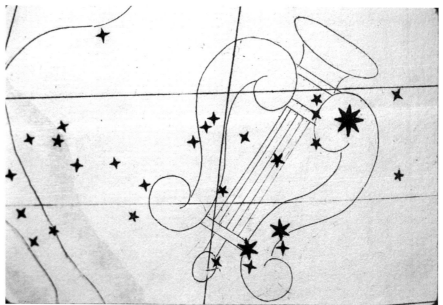

Lyra
Wollaston, Francis. A Portrature of the Heavens, as They Appear to the Naked Eye, *London, 1811.*

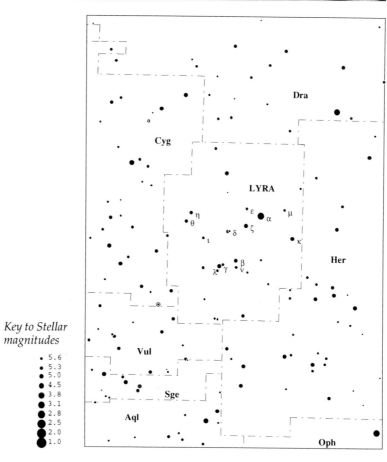

Key to Stellar magnitudes

. 5.6
. 5.3
. 5.0
• 4.5
• 3.8
• 3.1
● 2.8
● 2.5
● 2.0
● 1.0

Mensa

Meaning:	The Table Mountain
Pronunciation:	men' suh
Abbreviation:	Men
Possessive form:	Mensae (men' sigh)
Asterisms:	none
Bordering constellations:	Chamaeleon, Dorado, Hydrus, Octans, Volans
Overall brightness:	5.212 (73)
Central point:	RA = 5h28m Dec. = −77.5°
Directional extremes:	N = −70° S = −85° E = 7h37m W = 3h20m
Messier objects:	none
Meteor showers:	none
Midnight culmination date:	14 Dec
Bright stars:	none
Named stars:	none
Near stars:	α Men (136)
Size:	153.48 square degrees (0.372% of the sky)
Rank in size:	75
Solar conjunction date:	14 Jun
Visibility:	completely visible from latitudes: S of +5°
	completely invisible from latitudes: N of +20°
Visible stars:	(number of stars brighter than magnitude 5.5): 8
Interesting facts:	(1) This was one of the 14 constellations invented by Lacaille during his stay at the Cape of Good Hope in 1751–2.

Mensa (labeled 'Montagne de la Table' on this map) Lacaille, Nicolas Louis de. Planisphere contenant les Constellations Celestes, found in Mémoires Académie Royale des Sciences, Paris, 1752 (published in 1756). This constellation was invented by Lacaille and the photo shows its first appearance on any star map.

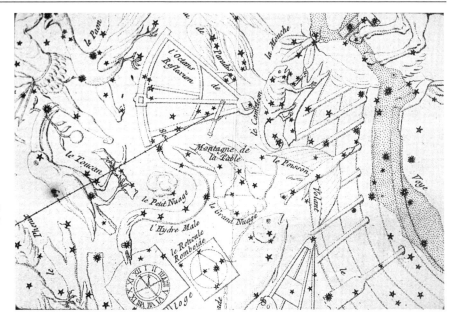

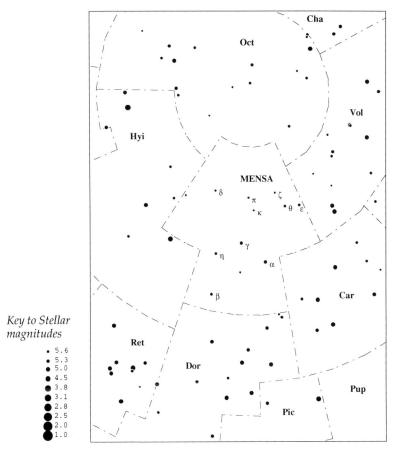

Key to Stellar magnitudes

Microscopium

Meaning:	The Microscope
Pronunciation:	my krow scop′ ee um
Abbreviation:	Mic
Possessive form:	Microscopii (my krow skow′ pee ee)
Asterisms:	none

Bordering constellations: Capricornus, Grus, Indus, Piscis Austrinus, Sagittarius

Overall brightness:	7.160 (42)
Central point:	RA = 20h55m Dec. = −36.5°

Directional extremes: N = −28°   S = −45°   E = 21h25m   W = 20h25m

Messier objects:	none
Meteor showers:	none

Midnight culmination date: 4 Aug

Bright stars:	none
Named stars:	none
Near stars:	Cordoba 29191 (21), LTT 8181-8182 (119), LTT 8214 (164)
Size:	209.51 square degrees (0.508% of the sky)
Rank in size:	66

Solar conjunction date: 4 Feb

Visibility:	completely visible from latitudes: S of +45°
	completely invisible from latitudes: N of +62°
Visible stars:	(number of stars brighter than magnitude 5.5): 15

Interesting facts: (1) This was one of the 14 constellations invented by Lacaille during his stay at the Cape of Good Hope in 1751–2.

Microscopium (labeled 'le Microscope' on this map)
Lacaille, Nicolas Louis de. Planisphere contenant les Constellations Celestes, found in Mémoires Académie Royale des Sciences, Paris, 1752 (published in 1756). This constellation was invented by Lacaille and the photo shows its first appearance on any star map.

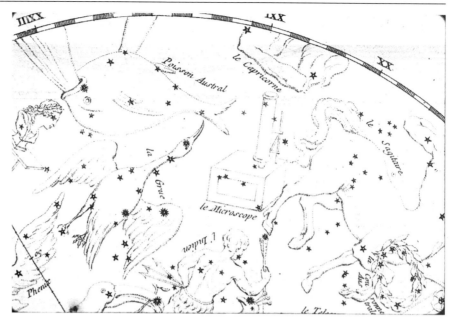

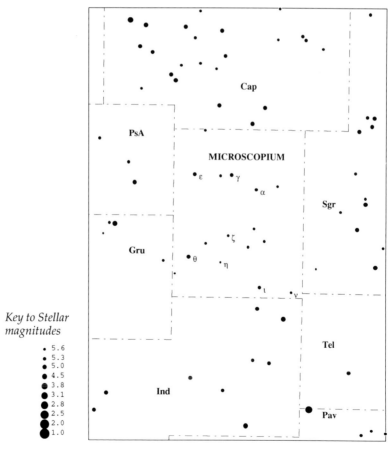

Key to Stellar magnitudes

- 5.6
- 5.3
- 5.0
- 4.5
- 3.8
- 3.1
- 2.8
- 2.5
- 2.0
- 1.0

Monoceros

Meaning:	The Unicorn
Pronunciation:	mon oss' sir us
Abbreviation:	Mon
Possessive form:	Monocerotis (mon awe sir awe' tiss)
Asterisms:	none

Bordering constellations: Canis Major, Canis Minor, Gemini, Hydra, Lepus, Orion, Puppis

Overall brightness:	7.476 (38)
Central point:	RA = 7h01m Dec. = +0.5°
Directional extremes:	N = +12° S = –11° E = 8h08m W = 5h54m
Messier objects:	M50
Meteor showers:	Monocerotids (10 Dec)

Midnight culmination date: 5 Jan

Bright stars:	none
Named stars:	none
Near stars:	Ross 614 A-B (24), BD–5°1844 A-B (171)
Size:	481.57 square degrees (1.167% of the sky)
Rank in size:	35

Solar conjunction date: 8 Jul

Visibility:	completely visible from latitudes: +79° to –78°
	portions visible worldwide
Visible stars:	(number of stars brighter than magnitude 5.5): 36

Interesting facts:
(1) This constellation first appeared in 1613, on a celestial globe designed by the Dutch theologian Petrus Plancius.

(2) One of the most unusual double stars in the sky resides in this constellation. Known as 'Plaskett's Star,' this is a pair of extremely massive stars, possibly the most massive pair yet identified. Its position is near the star 13 Mon, almost directly on the galactic equator. The total mass of this system is more than 100 times that of the Sun.

(3) One of the most beautiful of all galactic nebula is the Rosette Nebula, a faint ring of wispy material surrounding an open cluster containing the star 12 Mon. The cluster is designated NGC 2244, and the complex nebula's three brightest parts have been given the numbers NGC 2237, 2238, and 2239.

(4) Also in this constellation is one of the strangest of the nebulae. Dubbed 'Hubble's Variable Nebula,' this object not only changes its brightness, but its size and shape as well. The nebula surrounds the variable star R Mon, but the periods of the two objects do not seem to be related. No regular pattern of variability has been found for the nebula.

Monoceros
Rost, Johann
Leonhard. Atlas
Portatilis Coelestis,
Nuremburg, 1723.

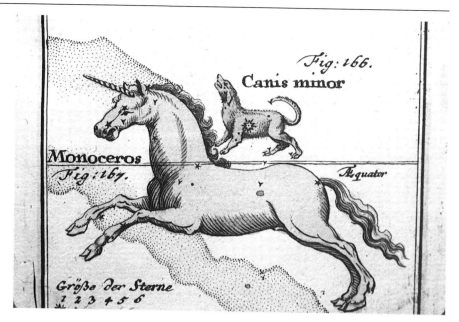

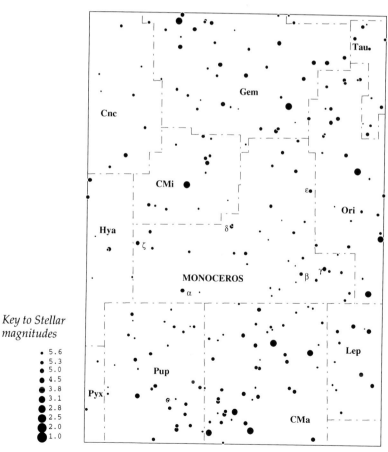

Key to Stellar
magnitudes

245

Musca

Meaning:	The Fly
Pronunciation:	mus′ kuh
Abbreviation:	Mus
Possessive form:	Muscae (mus′ kye)
Asterisms:	none
Bordering constellations:	Apus, Carina, Centaurus, Chamaeleon, Circinus, Crux
Overall brightness:	13.732 (8)
Central point:	RA = 12h31m Dec. = −69.5°
Directional extremes:	N = −64° S = −75° E = 13h46m W = 11h17m
Messier objects:	none
Meteor showers:	none
Midnight culmination date:	30 Mar
Bright stars:	α (110), β (179)
Named stars:	none
Near stars:	L 145-141 (36)
Size:	138.36 square degrees (0.335% of the sky)
Rank in size:	77
Solar conjunction date:	29 Sep
Visibility:	completely visible from latitudes: S of +15°
	completely invisible from latitudes: N of +26°
Visible stars:	(number of stars brighter than magnitude 5.5): 19
Interesting facts:	(1) This is one of 11 constellations invented by Pieter Dirksz Keyser and Frederick de Houtman, during the years 1595–7.

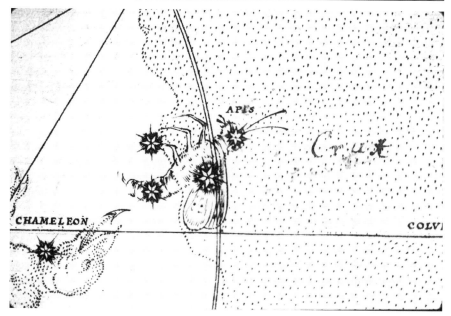

Musca (labeled 'Apis' on this map) Bayer, Johann. Uranometria, *Augsburg, 1603. This constellation was invented by de Houtman and Keyser in 1596. It was first illustrated on a globe by Plancius, which has not survived. This photo from Bayer's map, therefore, shows the earliest existing picture of this constellation.*

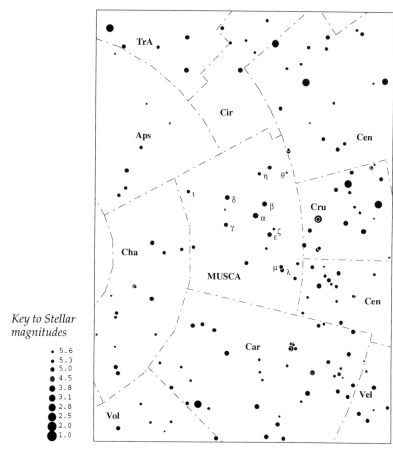

Key to Stellar magnitudes

· 5.6
· 5.3
· 5.0
● 4.5
● 3.8
● 3.1
● 2.8
● 2.5
● 2.0
● 1.0

Norma

Meaning: The Carpenter's Square

Pronunciation: nor' muh

Abbreviation: Nor

Possessive form: Normae (nor' mye)

Asterisms: none

Bordering constellations: Ara, Circinus, Lupus, Scorpius, Triangulum Australe

Overall brightness: 8.470 (31)

Central point: RA = 15h58m   Dec. = −51°

Directional extremes: N = −42°   S = −60°   E = 16h31m   W = 15h25m

Messier objects: none

Meteor showers: none

Midnight culmination date: 19 May

Bright stars: none

Named stars: none

Near stars: none

Size: 165.29 square degrees (0.401% of the sky)

Rank in size: 74

Solar conjunction date: 21 Nov

Visibility: completely visible from latitudes: S of +30°
completely invisible from latitudes: N of +48°

Visible stars: (number of stars brighter than magnitude 5.5): 14

Interesting facts: (1) This was one of the 14 constellations invented by Lacaille during his stay at the Cape of Good Hope in 1751–2.

Norma (labeled 'l'Equerre et la Regle' on this map) Lacaille, Nicolas Louis de. Planisphere contenant les Constellations Celestes, found in Mémoires Académie Royale des Sciences, Paris, 1752 (published in 1756). This constellation was invented by Lacaille and the photo shows its first appearance on any star map.

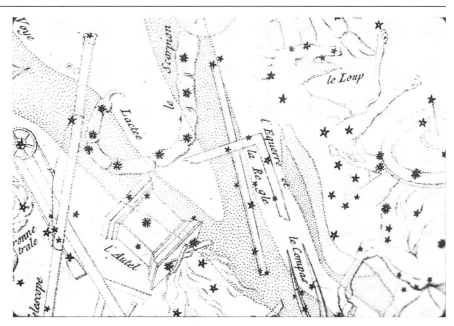

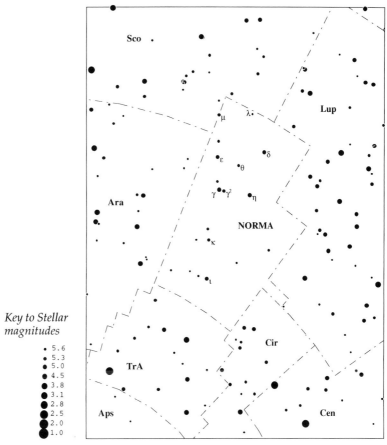

Key to Stellar magnitudes

249

Octans

Meaning:	The Octant
Pronunciation:	ok' tans
Abbreviation:	Oct
Possessive form:	Octantis (ok tan' tiss)
Asterisms:	none

Bordering constellations: Apus, Chamaeleon, Hydrus, Indus, Mensa, Pavo, Tucana

Overall brightness:	5.841 (60)
Central point:	RA = circumpolar Dec. = $-82.5°$
Directional extremes:	N = $-75°$ S = $-90°$ E = circumpolar W = circumpolar
Messier objects:	none
Meteor showers:	none

Midnight culmination date: none (circumpolar)

Bright stars:	none
Named stars:	none
Near stars:	LFT 1747 (120), LFT 1813 (183)
Size:	291.05 square degrees (0.706% of the sky)
Rank in size:	50

Solar conjunction date: none (circumpolar)

Visibility:	completely visible from latitudes: S of +00°
	completely invisible from latitudes: N of +15°
Visible stars:	(number of stars brighter than magnitude 5.5): 17

Interesting facts:
(1) This was one of the 14 constellations invented by Lacaille during his stay at the Cape of Good Hope in 1751–2.
(2) Octans is circumpolar, that is, it completely surrounds the south celestial pole. The nearest visible star to the actual location of the pole is σ Oct, at magnitude 5.47. Polaris (α UMi), the bright star closest to the north celestial pole, shines at magnitude 2.02, more than 22 times as brightly as σ Oct.

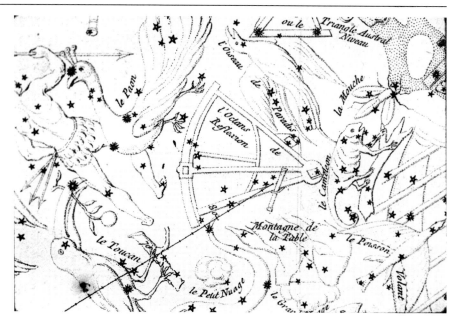

*Octans (labeled 'l'Octans Reflexion' on this map)
Lacaille, Nicolas Louis de. Planisphere contenant les Constellations Celestes, found in* Mémoires Académie Royale des Sciences, *Paris, 1752 (published in 1756). This constellation was invented by Lacaille and the photo shows its first appearance on any star map.*

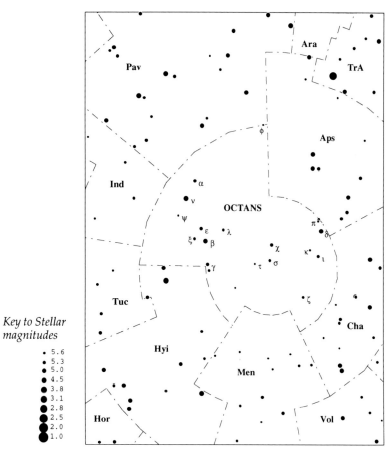

Key to Stellar magnitudes

- 5.6
- 5.3
- 5.0
- 4.5
- 3.8
- 3.1
- 2.8
- 2.5
- 2.0
- 1.0

Ophiuchus

Meaning:	The Serpent Bearer
Pronunciation:	off ee oo' kus
Abbreviation:	Oph
Possessive form:	Ophiuchi (off ee oo' key)
Asterisms:	The Bull of Poniatowski

Bordering constellations: Aquila, Hercules, Libra, Sagittarius, Scorpius, Serpens

Overall brightness:	5.800 (62)
Central point:	RA = 17h20m Dec. = –8°
Directional extremes:	N = +14° S = –30° E = 18h42m W = 15h58m
Messier objects:	M9, M10, M12, M14, M19, M62, M107
Meteor showers:	θ Ophiuchids (13 Jun)

Midnight culmination date: 11 Jun

Bright stars:	α (56), η (83), ζ (95), δ (115), β (122)
Named stars:	Cebalrai (β), Cheleb (β), Kelb Alrai (β), Rasalhague (α), Sabik (η), Yed Posterior (ε), Yed Prior (δ)
Near stars:	Barnard's Star (3), Wolf 1061 (25), 70 Oph A-B (42), 36 Oph A-B (45), LFT 1332 (47), Wolf 629 (62), V1054 Oph A-B-C (72), Wolf 718 (99), Wolf 751 (176), Wolf 636 (190)
Size:	948.34 square degrees (2.299% of the sky)
Rank in size:	11

Solar conjunction date: 12 Dec

Visibility:	completely visible from latitudes: +60° to –76°
	portions visible worldwide
Visible stars:	(number of stars brighter than magnitude 5.5): 55

Interesting facts: (1) Within the constellation of Ophiuchus lies Barnard's Star, the star with the greatest proper motion of any in the sky. This 'runaway star,' as such stars were called in the last century, moves across our field of view at the rate of 10.29 seconds of arc per year. This means that in only 175 years, Barnard's Star will have changed its position by the width of the Moon! This is also a very nearby star. In fact, after the α Cen system, it is the nearest star to the Earth, lying at a distance of only 5.95 light years. Barnard's Star is a red dwarf star shining at an apparent magnitude of 9.5. Irregularities in its motion suggest to some the possible existence of planetary bodies in orbit around this star.

(2) The most recent great supernova explosion in our Milky Way galaxy was observed in this constellation in the year 1604. It is known as 'Kepler's Nova,' due to the detailed study this astronomer made of this object, although Kepler was not the first to observe it. It was unusual that on the date the supernova first appeared, it was only 3° to the northwest of Mars and Jupiter, which were in conjunction, and only 4° to the east of Saturn. At maximum brightness, approximately three weeks after its discovery, this object shone at an estimated visual magnitude of –3.

Ophiuchus
Semler, Christoph.
Coelum Stellatum in
quo asterisimi...,
Halle, 1731.

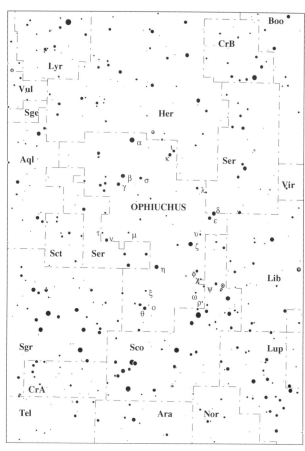

*Key to Stellar
magnitudes*

253

Orion

Meaning:	The Hunter
Pronunciation:	or eye' on
Abbreviation:	Ori
Possessive form:	Orionis (or ee oh' niss)
Asterisms:	The Belt, The Butterfly, The Heavenly G, The Rake, The Sword, The Three Kings, Venus' Mirror, The Winter Octagon, The Winter Oval, The Winter Triangle

Bordering constellations: Eridanus, Gemini, Lepus, Monoceros, Taurus

Overall brightness:	12.960 (9)
Central point:	RA = 5h32m Dec. = +6°
Directional extremes:	N = +23° S = −11° E = 6h23m W = 4h41m
Messier objects:	M42, M43, M78
Meteor showers:	Orionids (21 Oct)
	S. χ Orionids (10 Dec)
	N. χ Orionids (11 Dec)

Midnight culmination date: 13 Dec

Bright stars:	β (7), α (10), γ (26), ε (29), ζ (31), κ (53), δ (63), ι (121)
Named stars:	Algebar (β), Alnilam (ε), Alnitak (ζ), Bellatrix (γ), Betelgeuse (α), Hatsya (ι), Heka (λ), Meissa (λ), Mintaka (δ), Rigel (β), Saiph (κ), Tabit (π³)
Near stars:	Wolf 1453 (59), Ross 47 (65), LP 658–2 (66), π³ Ori (98), Ross 41 (135), BD+10°1032 A-B (169), χ¹ Ori (188)
Size:	594.12 square degrees (1.440% of the sky)
Rank in size:	26

Solar conjunction date: 15 Jun

Visibility:	completely visible from latitudes: +79° to −67°
	portions visible worldwide
Visible stars:	(number of stars brighter than magnitude 5.5): 77

Interesting facts:

(1) α Ori, or Betelgeuse, is the only first-magnitude star which is also variable. Its brightness changes in an irregular fashion from a minimum of 1.3 to a maximum of about 0.4.

(2) β Ori, or Rigel, is one of the most luminous stars in the sky. It is a blue supergiant with an absolute magnitude estimated at −7.1. This makes Rigel 47 863 times as bright as our Sun.

(3) In the region around the star ζ Ori lies an area of nebulosity containing one of the most famous – and one of the most visually elusive – objects in the sky. This is the Horsehead Nebula, a region of dark matter silhouetted against a brightly lit cloud of interstellar gas. Photographs taken with large telescopes show this object in great detail, but even the best amateur telescopes are sorely tested to resolve the Horsehead.

(4) M42, the Orion Nebula, is the best known and most visually structured diffuse nebula in the sky. This is a region of star formation, and several recently formed stars may be glimpsed enmeshed within the nebulosity. M42 lies at an approximate distance of 1500 light years. It was the first nebula to be photographed. A picture of it was taken in 1880 by Henry Draper.

Orion
Piccolomini,
Alessandro. De la
Sfera del Mondo...;
De le Stelle Fisse,
Venice, 1540.

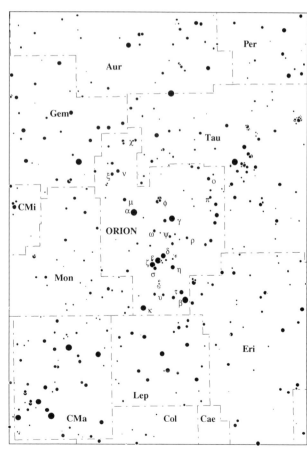

Key to Stellar
magnitudes

255

Pavo

Meaning:	The Peacock
Pronunciation:	pah' voe
Abbreviation:	Pav
Possessive form:	Pavonis (puh voe' niss)
Asterisms:	none

Bordering constellations: Apus, Ara, Indus, Octans, Telescopium

Overall brightness: 7.414 (39)

Central point: RA = 19h33m   Dec. = −66°

Directional extremes: N = −57°   S = −75°   E = 21h30m   W = 17h37m

Messier objects:	none
Meteor showers:	none

Midnight culmination date: 15 Jul

Bright stars:	α (44)
Named stars:	Peacock (α)
Near stars:	δ Pav (51), LFT 1372 (58), γ Pav (134)
Size:	377.67 square degrees (0.916% of the sky)
Rank in size:	44

Solar conjunction date: 14 Jan

Visibility: completely visible from latitudes: S of +15°
completely invisible from latitutdes: N of +33°

Visible stars: (number of stars brighter than magnitude 5.5): 28

Interesting facts: (1) This is one of 11 constellations invented by Pieter Dirksz Keyser and Frederick de Houtman, during the years 1595–7.

Pavo
Bayer, Johann. Uranometria, *Augsburg, 1603. This constellation was invented by de Houtman and Keyser in 1596. It was first illustrated on a globe by Plancius, which has not survived. This photo from Bayer's map, therefore, shows the earliest existing picture of this constellation.*

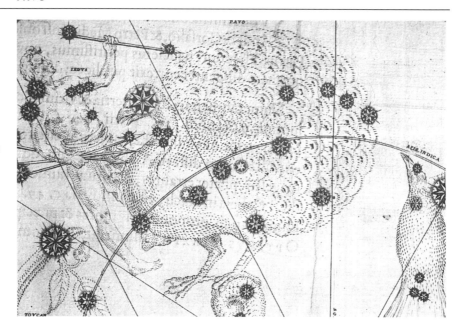

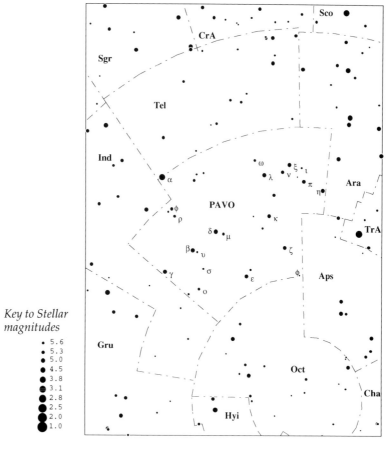

Key to Stellar magnitudes

257

Pegasus

Meaning:	The Winged Horse
Pronunciation:	peg′ ah sus
Abbreviation:	Peg
Possessive form:	Pegasi (peg′ uh see)
Asterisms:	The Baseball Diamond, The Great Square, The Large Dipper

Bordering constellations: Andromeda, Aquarius, Cygnus, Delphinus, Equuleus, Lacerta, Pisces, Vulpecula

Overall brightness:	5.086 (75)
Central point:	RA = 22h39m Dec. = +19°

Directional extremes: N = +36°   S = +2°   E = 00h13m   W = 21h06m

Messier objects:	M15
Meteor showers:	ξ Pegasids (9 Jul)
	Pegasids (12 Nov)

Midnight culmination date: 1 Sep

Bright stars:	ε (80), β (82), α (89), γ (132), η (155)
Named stars:	Algenib (γ), Baham (θ), Enif (ε), Homam (ζ), Markab (α), Matar (η), Sadalbari (μ), Scheat (β)
Near stars:	EQ Peg A-B (73), Ross 775 (74), Ross 671 (83)
Size:	1120.79 square degrees (2.717% of the sky)
Rank in size:	7

Solar conjunction date: 2 Mar

Visibility:	completely visible from latitudes: N of –54°
	completely invisible from latitudes: S of –88°
Visible stars:	(number of stars brighter than magnitude 5.5): 57

Interesting facts:
(1) The first object in J. L. E. Dreyer's *New General Catalog*, NGC 1, lies within the boundaries of this constellation. Its position is about 2° south and slightly west of α And, near the star 85 Peg. NGC 1 is a relatively faint 13th magnitude galaxy which shows little detail in amateur instruments.

(2) The asterism of the 'Great Square of Pegasus' is easily seen. Although this square appears to encompass an area bereft of bright stars, a good test of vision, and of the darkness of one's observing site, is to count the number of visible stars within the area of the Great Square. Observers have reported seeing between 30–50 stars in this region of sky.

(3) The star Alpheratz (α And) is a good example of what was once called a 'shared star.' On some ancient maps, this star appeared as δ Peg, thus completing the 'Great Square.' In 1928, however, the IAU permanently assigned this star to Andromeda, thus ending its 'shared' status.

Pegasus
Piccolomini,
Alessandro. De la
Sfera del Mondo...;
De le Stelle Fisse,
Venice, 1540.

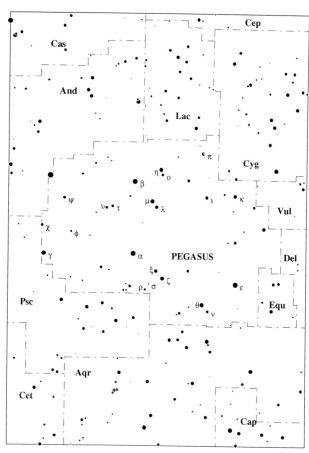

Key to Stellar
magnitudes

- 5.6
- 5.3
- 5.0
- 4.5
- 3.8
- 3.1
- 2.8
- 2.5
- 2.0
- 1.0

Perseus

Meaning:	The Hero
Pronunciation:	pur' see us
Abbreviation:	Per
Possessive form:	Persei (per' see ee)
Asterisms:	The Large Dipper, The Segment

Bordering constellations: Andromeda, Aries, Auriga, Camelopardalis, Cassiopeia, Taurus, Triangulum

Overall brightness:	10.569 (17)
Central point:	RA = 3h06m Dec. = +45°

Directional extremes: N = +59° S = +31° E = 4h46m W = 1h26m

Messier objects:	M34, M76
Meteor showers:	Daytime ζ Perseids (7 Jun)
	Perseids (12 Aug)

Midnight culmination date: 7 Nov

Bright stars:	α (33), β (58), ζ (135), ε (144), γ (153), δ (170)
Named stars:	Algol (β), Atik (o), Menkib (ξ), Miram (ν), Mirfak (α), Misam (κ)
Near stars:	Ross 594 (197)
Size:	615.00 square degrees (1.491% of the sky)
Rank in size:	24

Solar conjunction date: 9 May

Visibility:	completely visible from latitudes: N of −31°
	completely invisible from latitudes: S of −59°
Visible stars:	(number of stars brighter than magnitude 5.5): 65

Interesting facts:

(1) The most famous of the visible stars of Perseus is Algol (β). It is an eclipsing binary star – the first such star discovered. Many have speculated that the variability of this star was known in ancient times, and was the reason for its name, 'Demon Star.' There is absolutely no proof which supports such ideas, however.

(2) The open clusters h and χ Persei (NGC 869 and NGC 884) are located between the head of Perseus and the 'W' of Cassiopeia. They are beautiful examples of galactic clusters and are are well resolved in small and medium aperture telescopes using low power eyepieces.

(3) In studies of the spiral structure of our Milky Way galaxy, astronomers have optically mapped three sections of 'arms' which make up the spiral shape. These are the Orion arm, the Sagittarius arm, and the Perseus arm. The Orion arm contains our solar system, the Sagittarius arm is closer to the center of the Milky Way, and the Perseus arm is further out. About 6000 light years of each 'arm' has been mapped.

(4) The most famous meteor shower of the year occurs within the boundaries of the constellation of Perseus. The Perseid meteor shower peaks on 12 August with a rate of about 60 meteors seen each hour. Traces of the shower can be detected several weeks before and after maximum. The meteors are fast and many leave characteristic 'smoke trails.' This shower is associated with comet Swift–Tuttle (1862 III).

*Perseus
Gallucci, Giovanni
Paolo.* Theatrum
Mundi, et Temporis,
Venice, 1588.

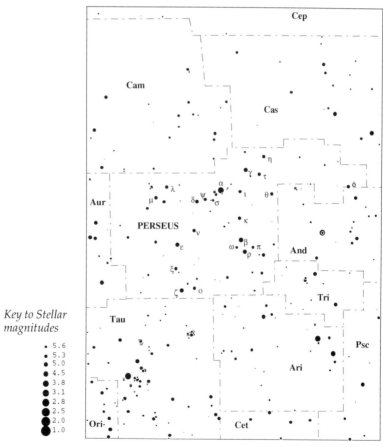

*Key to Stellar
magnitudes*

261

Phoenix

Meaning:	The Phoenix
Pronunciation:	fee′ niks
Abbreviation:	Phe
Possessive form:	Phoenicis (fen ee′ siss)
Asterisms:	none

Bordering constellations: Eridanus, Fornax, Grus, Sculptor, Tucana

Overall brightness:	5.753 (64)
Central point:	RA = 00h54m Dec. = −49°

Directional extremes: N = −40°   S = −58°   E = 2h24m   W = 23h24m

Messier objects:	none
Meteor showers:	July Phoenicids (14 Jul)
	December Phoenicids (5 Dec)

Midnight culmination date: 4 Oct

Bright stars:	α (79)
Named stars:	Ankaa (α)
Near stars:	L 362-81 (121)
Size:	469.32 square degrees (1.138% of the sky)
Rank in size:	37

Solar conjunction date: 5 Apr

Visibility:	completely visible from latitudes: S of +32°
	completely invisible from latitudes: N of +50°
Visible stars:	(number of stars brighter than magnitude 5.5): 27

Interesting facts: (1) This is one of 11 constellations invented by Pieter Dirksz Keyser and Frederick de Houtman, during the years 1595–7.

Phoenix
Bayer, Johann.
Uranometria,
*Augsburg, 1603. This
constellation was
invented by de
Houtman and Keyser
in 1596. It was first
illustrated on a globe
by Plancius, which has
not survived. This
photo from Bayer's
map, therefore, shows
the earliest existing
picture of this
constellation.*

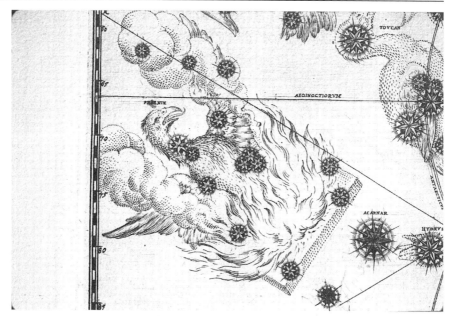

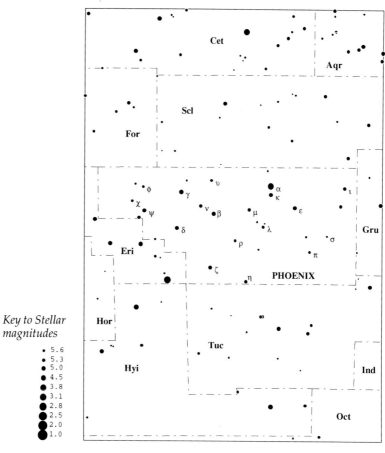

*Key to Stellar
magnitudes*

- 5.6
- 5.3
- 5.0
- 4.5
- 3.8
- 3.1
- 2.8
- 2.5
- 2.0
- 1.0

263

Pictor

Meaning:	The Painter's Easel
Pronunciation:	pik' tor
Abbreviation:	Pic
Possessive form:	Pictoris (pik tor' iss)
Asterisms:	none

Bordering constellations: Caelum, Carina, Columba, Dorado, Puppis, Volans

Overall brightness:	6.080 (55)
Central point:	RA = 5h41m Dec. = –53.5°

Directional extremes: N = –43°   S= –64°   E = 6h51m   W = 4h32m

Messier objects:	none
Meteor showers:	none

Midnight culmination date: 16 Dec

Bright stars:	none
Named stars:	none
Near stars:	Kapteyn's Star (22)
Size:	246.73 square degrees (0.598% of the sky)
Rank in size:	59

Solar conjunction date: 17 Jun

Visibility:	completely visible from latitudes: S of +26°
	completely invisible from latitudes: N of +47°
Visible stars:	(number of stars brighter than magnitude 5.5): 15

Interesting facts: (1) This was one of the 14 constellations invented by Lacaille during his stay at the Cape of Good Hope in 1751–2.

(2) In recent images taken of the star β Pic, astronomers have seen a disk of material surrounding the star. It is believed that this material represents direct proof of planetary formation, although no planets have as yet been seen.

Pictor (labeled 'le Chevalet et la Palette' on this map) Lacaille, Nicolas Louis de. Planisphere contenant les Constellations Celestes, found in Mémoires Académie Royale des Sciences, Paris, 1752 (published in 1756). This constellation was invented by Lacaille and the photo shows its first appearance on any star map.

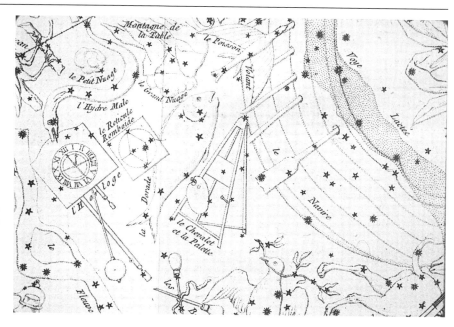

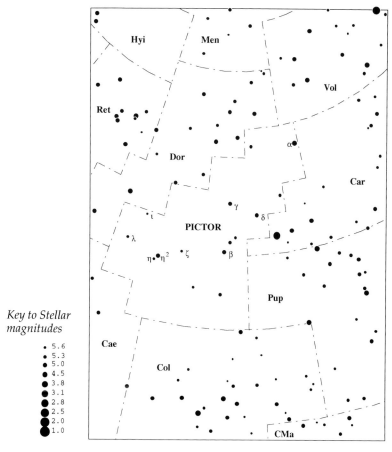

Key to Stellar magnitudes

- 5.6
- 5.3
- 5.0
- 4.5
- 3.8
- 3.1
- 2.8
- 2.5
- 2.0
- 1.0

265

Pisces

Meaning:	The Fishes
Pronunciation:	pie' seez
Abbreviation:	Psc
Possessive form:	Piscium (pish' ee um)
Asterisms:	The Circlet

Bordering constellations: Andromeda, Aquarius, Aries, Cetus, Pegasus, Triangulum

Overall brightness:	5.622 (68)
Central point:	RA = 00h26m Dec. = +13°

Directional extremes: N = +33°   S = –7°   E = 2h04m   W = 22h49m

Messier objects:	M74
Meteor showers:	S. Piscids (20 Sep)
	N. Piscids (12 Oct)

Midnight culmination date: 27 Sep

Bright stars:	none
Named stars:	Alrischa (α), Okda (α)
Near stars:	van Maanen's Star (26), BD+1°4774 (52), BD+4°123 (84), 107 Psc (95), LFT 1851 (139), 54 Psc (192)
Size:	889.42 square degrees (2.156% of the sky)
Rank in size:	14

Solar conjunction date: 29 Mar

Visibility:	completely visible from latitudes: +83° to –57°
	portions visible worldwide
Visible stars:	(number of stars brighter than magnitude 5.5): 50

Interesting facts: (1) The vernal equinox, that point at which the sun's right ascension and declination are both zero, is located in Pisces. Often referred to as the 'First Point of Aries,' (which was its location 2000 years ago when astrology was formalized) the vernal equinox has since migrated into Pisces due to the precessional motion of the Earth. In about 800 years, the vernal equinox will move into the constellation of Aquarius. The length of the entire precessional cycle is approximately 25 800 years.

Pisces
Flamsteed, John. Atlas
Coelestis, *London,*
1729.

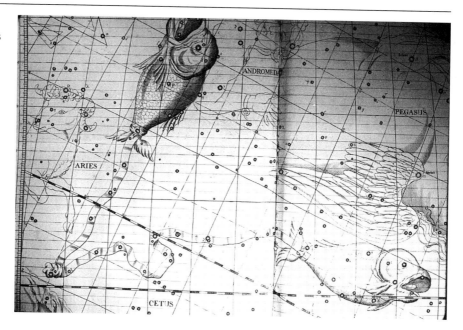

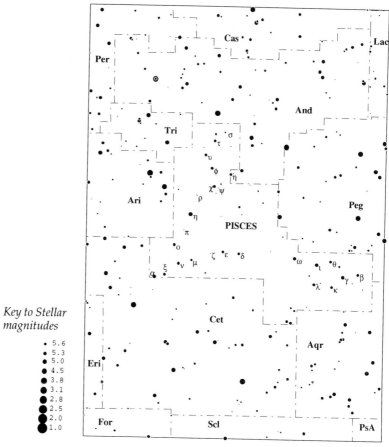

Key to Stellar
magnitudes

·	5.6
·	5.3
•	5.0
•	4.5
●	3.8
●	3.1
●	2.8
●	2.5
●	2.0
●	1.0

Piscis Austrinus

Meaning:	The Southern Fish
Pronunciation:	pie′ siss os try′ nus
Abbreviation:	PsA
Possessive form:	Piscis Austrini (pie′ siss os tree′ nee)
Asterisms:	none
Bordering constellations:	Aquarius, Capricornus, Grus, Microscopium, Sculptor
Overall brightness:	6.113 (54)
Central point:	RA = 22h14m Dec. = −31°
Directional extremes:	N = −25° S = −37° E= 23h04m W= 21h25m
Messier objects:	none
Meteor showers:	none
Midnight culmination date:	25 Aug
Bright stars:	α (18)
Named stars:	Fomalhaut (α)
Near stars:	Lacaille 9352 (19), α (80), LTT 9283 (108)
Size:	245.37 square degrees (0.595% of the sky)
Rank in size:	60
Solar conjunction date:	24 Feb
Visibility:	completely visible from latitudes: S of +53°
	completely invisible from latitudes: N of +65°
Visible stars:	(number of stars brighter than magnitude 5.5): 15
Interesting facts:	(1) α PsA, or Fomalhaut, is one of the four Royal Stars of the ancient Persians. The other three are Aldebaran (α Tau), Antares (α Sco), and Regulus (α Leo).

Piscis Austrinus
Burritt, Elijah H.
Atlas Designed to
Illustrate the
Geography of the
Heavens, *Hartford,*
1835.

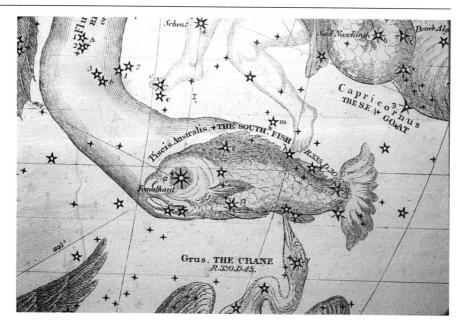

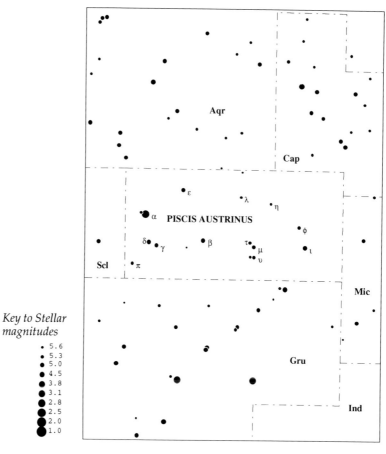

Key to Stellar
magnitudes

269

Puppis

Meaning:	The Stern (of Argo Navis)
Pronunciation:	pup′ iss
Abbreviation:	Pup
Possessive form:	Puppis (pup′ iss)
Asterisms:	none

Bordering constellations: Canis Major, Carina, Columba, Hydra, Monoceros, Pictor, Pyxis, Vela

Overall brightness: 13.810 (7)

Central point: RA = 7h14m   Dec. = −31°

Directional extremes: N = −11°   S = −51°   E = 8h26m   W = 6h02m

Messier objects: M46, M47, M93

Meteor showers: π Puppids (23 Apr)

Midnight culmination date: 8 Jan

Bright stars: ζ (67), π (112), ρ (128), τ (154)

Named stars: Asmidiske (ξ), Markeb (k^1), Naos (ζ)

Near stars: LFT 571 (57), L 745–46 A-B (86), LFT 502 A-B (137)

Size: 673.43 square degrees (1.633% of the sky)

Rank in size: 20

Solar conjunction date: 11 Jul

Visibility: completely visible from latitudes: S of +39°
completely invisible from latitudes: N of +79°

Visible stars: (number of stars brighter than magnitude 5.5): 93

Interesting facts: (1) One of three constellations into which Lacaille divided the ancient constellation of Argo Navis. The other two 'sub-constellations' are Carina and Vela.

Puppis (labeled 'la Pouppe' on this map) Vaugondy,Robert de. Hémisphère Céleste Antarctique..., *Paris, 1764. This constellation was invented by Lacaille and included in his star catalog, but not pictured on any of his star charts. This photo from Robert de Vaugondy's map, therefore, shows the first appearance of Puppis as a separate constellation.*

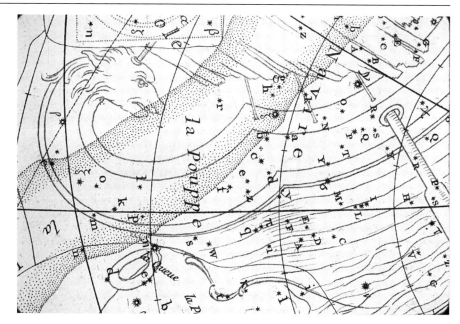

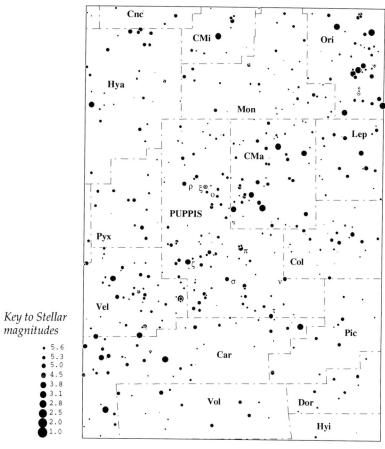

Key to Stellar magnitudes

5.6
5.3
5.0
4.5
3.8
3.1
2.8
2.5
2.0
1.0

Pyxis

Meaning:	The Compass (of Argo Navis)
Pronunciation:	pik' siss
Abbreviation:	Pyx
Possessive form:	Pyxidis (pik' si diss)
Asterisms:	none
Bordering constellations:	Antlia, Hydra, Puppis, Vela
Overall brightness:	5.434 (72)
Central point:	RA = 8h56m Dec. = –27°
Directional extremes:	N = –17° S = –37° E = 9h26m W = 8h26m
Messier objects:	none
Meteor showers:	none
Midnight culmination date:	4 Feb
Bright stars:	none
Named stars:	none
Near stars:	LFT 598 (131), L 532-81 (143)
Size:	220.83 square degrees (0.535% of the sky)
Rank in size:	65
Solar conjunction date:	6 Aug
Visibility:	completely visible from latitudes: S of +53°
	completely invisible from latitudes: N of +73°
Visible stars:	(number of stars brighter than magnitude 5.5): 12

Interesting facts: (1) This was one of the 14 constellations invented by Lacaille during his stay at the Cape of Good Hope in 1751–2. It is not one of the constellations into which he subdivided the ancient constellation of Argo Navis. In fact, on his celestial map of 1752, Lacaille pictures Argo Navis as a complete ship, with Pyxis just to its north.

(2) Within this constellation is a variable star designated T Pyx. This is the most active of all known recurring novae. Normally this star's magnitude is an inconspicuous 14. Every 18 to 24 years (the period varies) the brightness increases by as much as 1000 times, as the visual magnitude changes to 6.5.

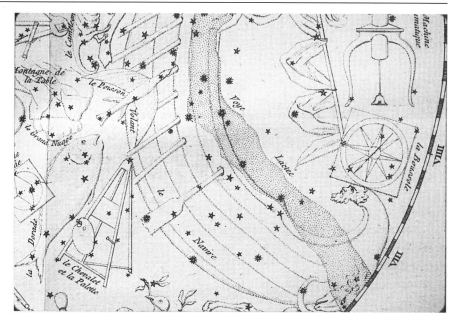

Pyxis (labeled 'la Boussole' on this map) Lacaille, Nicolas Louis de. Planisphere contenant les Constellations Celestes, found in Mémoires Académie Royale des Sciences, *Paris, 1752 (published in 1756). This constellation was invented by Lacaille and the photo shows its first appearance on any star map.*

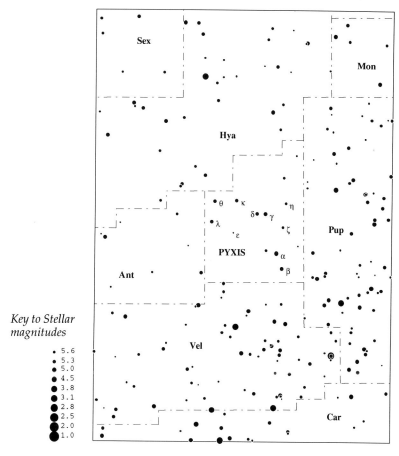

Key to Stellar magnitudes

273

Reticulum

Meaning:	The Net
Pronunciation:	reh tik′ yoo lum
Abbreviation:	Ret
Possessive form:	Reticuli (reh tik′ yoo lee)
Asterisms:	none

Bordering constellations: Dorado, Horologium, Hydrus

Overall brightness:	9.654 (24)
Central point:	RA = 3h54m Dec. = −60°

Directional extremes: N = −53°   S = −67°   E = 4h35m   W = 3h14m

Messier objects:	none
Meteor showers:	none

Midnight culmination date: 19 Nov

Bright stars:	none
Named stars:	none
Near stars:	none
Size:	113.94 square degrees (0.276% of the sky)
Rank in size:	82

Solar conjunction date: 21 May

Visibility:	completely visible from latitudes: S of +23°
	completely invisible from latitudes: N of +37°
Visible stars:	(number of stars brighter than magnitude 5.5): 11

Interesting facts: (1) This was one of the 14 constellations invented by Lacaille during his stay at the Cape of Good Hope in 1751–2.

(2) The star ζ Ret was the subject of a much-publicized 'UFO' incident. Supposedly, on 19 September 1961, a New Hampshire couple named Betty and Barney Hill were abducted by aliens who showed Mrs. Hill a star map. Further study seemed to indicate that the area revealed by the map was part of the local solar neighborhood, and that the aliens came from the double star system ζ Reticuli.

Reticulum (labeled 'le Reticule Romboide' on this map) Lacaille, Nicolas Louis de. Planisphere contenant les Constellations Celestes, found in Mémoires Académie Royale des Sciences, *Paris, 1752 (published in 1756). This constellation was invented by Lacaille and the photo shows its first appearance on any star map.*

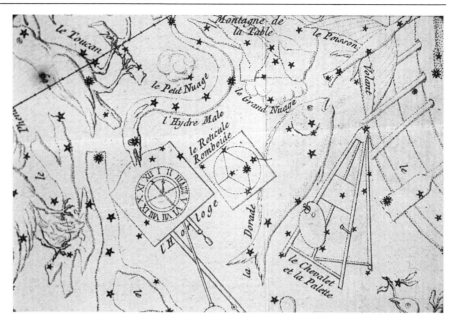

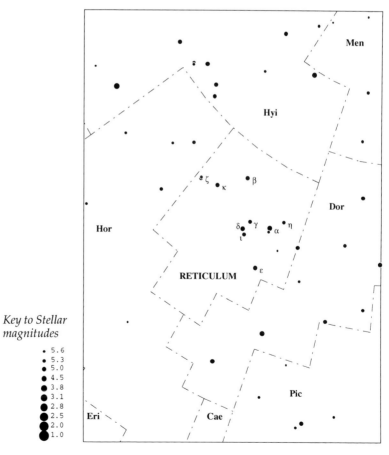

Key to Stellar magnitudes

. 5.6
. 5.3
. 5.0
. 4.5
● 3.8
● 3.1
● 2.8
● 2.5
● 2.0
● 1.0

Sagitta

Meaning:	The Arrow
Pronunciation:	suh gee' tuh
Abbreviation:	Sge
Possessive form:	Sagittae (suh jeet' eye)
Asterisms:	none

Bordering constellations: Aquila, Delphinus, Hercules, Vulpecula

Overall brightness: 10.009 (18)

Central point: RA = 19h37m   Dec. = +18.5°

Directional extremes: N = +21°   S= +16°   E= 20h18m   W= 18h56m

Messier objects:	M71
Meteor showers:	none

Midnight culmination date: 16 Jul

Bright stars:	none
Named stars:	Sham (α)
Near stars:	Ross 730-731 (128)
Size:	79.93 square degrees (0.194% of the sky)
Rank in size:	86

Solar conjunction date: 15 Jan

Visibility: completely visible from latitudes: N of –69°
completely invisible from latitudes: S of –74°

Visible stars: number of stars brighter than magnitude 5.5): 8

Interesting facts: (1) Sagitta is one of two constellations whose abbreviation (Sge) contains a letter ('e') not found in the constellation name. The other is Hydrus.

Sagitta
Bevis, John.
Uranographia
Britannica, *London,*
1745.

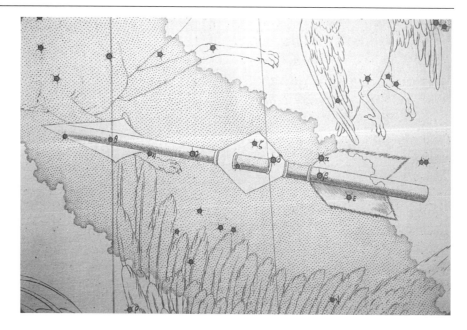

*Key to Stellar
magnitudes*

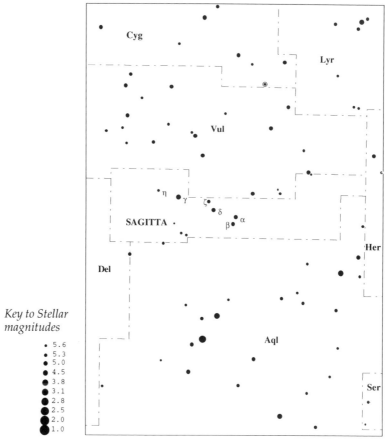

Sagittarius

Meaning:	The Archer
Pronunciation:	sa ji tare' ee us
Abbreviation:	Sgr
Possessive form:	Sagittarii (sa jit air' ee ee)
Asterisms:	The Milk Dipper
Bordering constellations:	Aquila, Capricornus, Corona Australis, Microscopium, Ophiuchus, Scorpius, Scutum, Serpens, Telescopium
Overall brightness:	7.493 (35)
Central point:	RA = 19h03m Dec. = −28.5°
Directional extremes:	N = −12° S = −45° E = 20h25m W = 17h41m
Messier objects:	M8, M17, M18, M20, M21, M22, M23, M24, M25, M28, M54, M55, M69, M70, M75
Meteor showers:	Sagittariids (11 Jun)
Midnight culmination date:	7 Jul
Bright stars:	ε (36), σ (49), ζ (99), δ (113), λ (130), π (148), γ (164), η (188)
Named stars:	Al Nasl (γ), Alrami (α), Arkeb Posterior (β²), Arkeb Prior (β¹), Ascella (ζ), Kaus Australis (ε), Kaus Boraelis (λ), Kaus Meridionalis (δ), Nash (γ), Nunki (σ), Nushaba (γ), Rukbat (α)
Near stars:	Ross 154 (8), LFT 1529-1530 (48), LFT 1469 (50), LFT 1532 (68)
Size:	867.43 square degrees (2.103% of the sky)
Rank in size:	15
Solar conjunction date:	6 Jan
Visibility:	completely visible from latitudes: S of +45°
	completely invisible from latitudes: N of +78°
Visible stars:	(number of stars brighter than magnitude 5.5): 65

Non-traditional 'mythology': The stars λ, φ, σ, τ, ζ, ε, γ, and δ of this constellation have long been known as the 'teapot' of Sagittarius, a figure, it must be noted, that is much easier to see than an archer who is half man, half horse. Along with the teapot in the sky are some related figures: a sugar spoon formed of the stars υ, ρ¹, π, ο, ν¹ and ν², and ξ² of Sagittarius; a slice of lemon (the constellation of Corona Australis); and steam rising from the spout of the teapot (the Milky Way).

Interesting facts:
(1) Sagittarius contains the greatest number of Messier objects of any constellation, a total of 15. (Second place goes to Virgo with 11.)
(2) The center of our galaxy (Galactic Latitude and Longitude = 0°) is within the boundaries of Sagittarius, at RA = 17h42m Dec = −29°. This point lies just to the southwest of the star × Sgr.
(3) Certainly the 'teapot' is the most recognizable feature of this constellation. It is interesting that α Sgr and β Sgr, not nearly the brightest stars of Sagittarius, are far to the southeast of the teapot, with Corona Australis lying in the space between.
(4) As most observers concentrate on the teapot, it may be surprising to discover that Sagittarius is quite a bit larger (by 75%!) than neighboring Scorpius.
(5) There are more designated variable stars in Sagittarius than there are naked-eye stars in the entire sky.

Sagittarius
Bode, Johann Elert.
Uranographia Sive
Astrorum
Descriptio, *Berlin,*
1801.

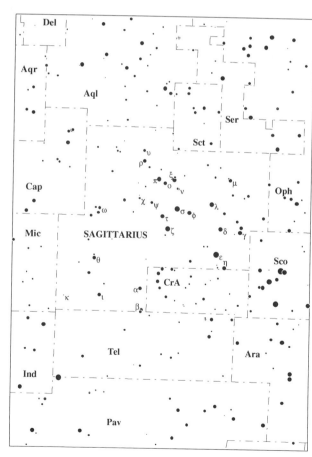

Key to Stellar
magnitudes

- 5.6
- 5.3
- 5.0
- 4.5
- 3.8
- 3.1
- 2.8
- 2.5
- 2.0
- 1.0

Scorpius

Meaning:	The Scorpion
Pronunciation:	skor' pee us
Abbreviation:	Sco
Possessive form:	Scorpii (skor' pee ee)
Asterisms:	The Fish Hook

Bordering constellations: Ara, Corona Australis, Libra, Lupus, Norma, Ophiuchus, Sagittarius

Overall brightness:	12.480 (10)
Central point:	RA = 16h49m Dec. = –27°
Directional extremes:	N = –8° S = –46° E = 17h55m W = 15h44m
Messier objects:	M4, M6, M7, M80
Meteor showers:	α Scorpiids (3 May)
	χ Scorpiids (5 Jun)

Midnight culmination date: 3 Jun

Bright stars:	α (15), λ (25), θ (39), ε (72), δ (76), κ (81), β (93), υ (111), τ (129), π (146), σ (147), ι^1 (176), μ^1 (182)
Named stars:	Acrab (β), Al Niyat (σ), Al Niyat (τ), Antares (α), Dschubba (δ), Graffias (ζ), Jabbah (ν), Lesath (ν), Sargas (θ), Shaula (λ), Vespertilio (α)
Near stars:	LFT 1358 (33), HD 156384 A-B-C (88), LFT 1266-1267 (102)
Size:	496.78 square degrees (1.204% of the sky)
Rank in size:	33

Solar conjunction date: 4 Dec

Visibility:	completely visible from latitudes: S of +44°
	completely invisible from latitudes: N of +82°
Visible stars:	(number of stars brighter than magnitude 5.5): 62

Interesting facts:

(1) In ancient times, Scorpius also contained the stars of the present day constellation of Libra, the Scales. Libra represented the claws of the Scorpion.

(2) α Sco, or Antares, is one of the four Royal Stars of the ancient Persians. The other three are Aldebaran (α Tau), Regulus (α Leo), and Fomalhaut (α PsA).

(3) α Sco is a reddish star, and since it lies in the band of the zodiac, the planets are often seen nearby. It was because of the frequent proximity of Mars, that this star received its name, 'Antares.' This title is a literal combination of the terms 'anti' and 'Ares,' meaning, of course, 'the rival of Mars.'

(4) About 5° NNE of ν Sco lies the strongest x-ray source in the sky. Designated Scorpius X-1, this object is a close binary star with an apparent magnitude of 13. One of the stars is probably a neutron star of high density. As gas enveloping the system streams into the intense gravitational and magnetic fields near this star, it is accelerated to speeds near that of light. The result is the emission of x-rays called synchrotron radiation.

Scorpius
Piccolomini,
Alessandro. De la
Sfera del Mondo...;
De le Stelle Fisse,
Venice, 1540.

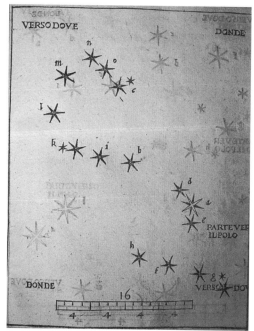

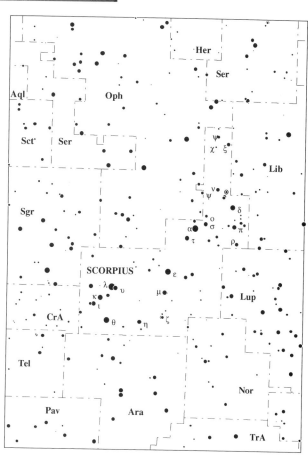

Key to Stellar
magnitudes

281

Sculptor

Meaning:	The Sculptor's Workshop
Pronunciation:	skulp' tor
Abbreviation:	Scl
Possessive form:	Sculptoris (skulp tor' iss)
Asterisms:	none

Bordering constellations: Aquarius, Cetus, Fornax, Grus, Phoenix, Piscis Austrinus

Overall brightness:	3.159 (86)
Central point:	RA = 00h24m Dec. = −32.5°
Directional extremes:	N = −25° S = −40° E = 1h44m W= 23h04m
Messier objects:	none
Meteor showers:	none

Midnight culmination date: 26 Sep

Bright stars:	none
Named stars:	none
Near stars:	BD+44°4548 (28)
Size:	474.76 square degrees (1.151% of the sky)
Rank in size:	36

Solar conjunction date: 29 Mar

Visibility:	completely visible from latitudes: S of +50°
	completely invisible from latitudes: N of +65°
Visible stars:	(number of stars brighter than magnitude 5.5): 15

Interesting facts: (1) This was one of the 14 constellations invented by Lacaille during his stay at the Cape of Good Hope in 1751–2.

(2) The south pole of the Milky Way galaxy lies within the borders of Sculptor. The coordinates of this point are approximately RA = 0h49m Dec. = −27.5°.

Sculptor (labeled 'l'Atelier du Sculpteur' on this map)
Lacaille, Nicolas Louis de. Planisphere contenant les Constellations Celestes, found in Mémoires Académie Royale des Sciences, *Paris, 1752 (published in 1756). This constellation was invented by Lacaille and the photo shows its first appearance on any star map.*

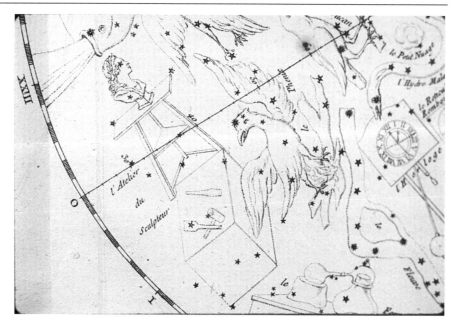

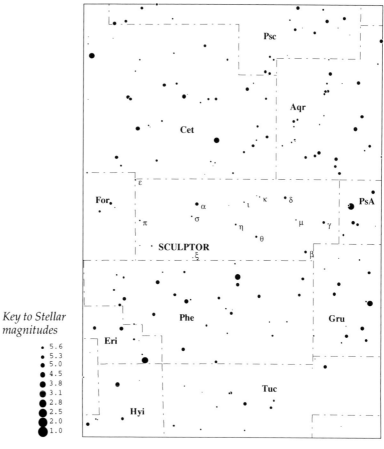

Key to Stellar magnitudes

283

Scutum

Meaning:	The Shield
Pronunciation:	skoo' tum
Abbreviation:	Sct
Possessive form:	Scuti (skoo'tee)
Asterisms:	none
Bordering constellations:	Aquila, Sagittarius, Serpens
Overall brightness:	8.249 (33)
Central point:	RA = 18h37m Dec. = −10°
Directional extremes:	N = −4° S = −16° E = 18h56m W = 18h18m
Messier objects:	M11, M26
Meteor showers:	none
Midnight culmination date:	1 Jul
Bright stars:	none
Named stars:	none
Near stars:	none
Size:	109.11 square degrees (0.265% of the sky)
Rank in size:	84
Solar conjunction date:	31 Dec
Visibility:	completely visible from latitudes: S of +74°
	completely invisible from latitudes: N of +86°
Visible stars:	(number of stars brighter than magnitude 5.5): 9

Interesting facts: (1) One of seven constellations still in use invented by Johannes Hevelius. In 1690, this group was included in a star atlas which accompanied his stellar catalog.

(2) Within the boundaries of this constellation lies one of the finest of all open, or galactic, clusters – M11. The common name of this object comes from the famous *Bedford Catalogue*, printed in 1844, by Captain (later Admiral) William Henry Smyth. The actual quote is as follows: 'This object, which somewhat resembles a flight of wild ducks in shape, is a gathering of minute stars, with a prominent 8th magnitude in the middle, and two following. . .' This object has been called the 'Wild Duck' ever since. This wondrous object can easily be seen through medium-sized (20 cm) telescopes, and is the personal favorite of this writer.

Scutum
Hevelius, Johannes.
Firmamentum
Sobiescianum, sive
Uranographia,
totum Coelum
Stellatum, *Gdansk,*
1690. This is the first
appearance of this
constellation on any
star map.

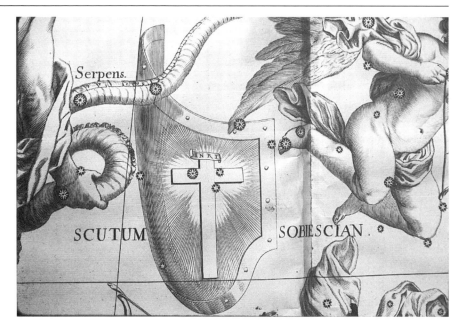

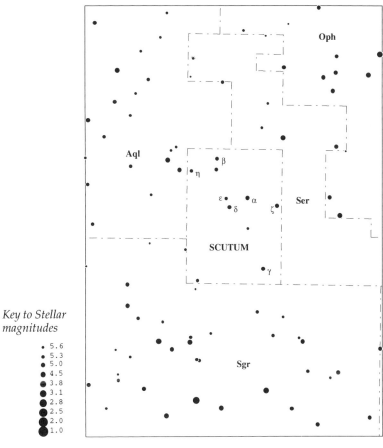

Key to Stellar
magnitudes

285

Serpens

Meaning:	The Serpent
Pronunciation:	sir′ pens
Abbreviation:	Ser
Possessive form:	Serpentis (sir pen′ tiss)
Asterisms:	none
Bordering constellations:	Aquila, Boötes, Corona Borealis, Hercules, Libra, Ophiuchus, Sagittarius, Scutum, Virgo
Overall brightness:	5.652 (67)
Central point:	RA = 16h55m Dec. = +5°
Directional extremes:	N = +26° S = –16° E = 18h56m W = 14h55m
Messier objects:	M5, M16
Meteor showers:	none
Midnight culmination date:	6 Jun
Bright stars:	α (104)
Named stars:	Alya (θ), Cor Serpentis (α), Unuk al Hai (α)
Near stars:	BD-3°4233 (93)
Size:	636.92 square degrees (1.544% of the sky)
Rank in size:	23
Solar conjunction date:	5 Dec
Visibility:	completely visible from latitudes: +74° to –64° portions visible worldwide
Visible stars:	(number of stars brighter than magnitude 5.5): 36

Interesting facts: (1) Approximately 10° directly north of β Lib lies the magnificent globular cluster M5. Perhaps, in the entire northern sky, only M13 in Hercules is more wonderful.

(2) In this constellation is another noteworthy object on Messier's list, M16, the famous Eagle Nebula. Somewhat disappointing in small and medium-sized instruments, on long-exposure photographs, a magnificent diffuse nebula is revealed. Robert Burnham, Jr., in his famous *Celestial Handbook*, Dover, 1978 christened this object the 'Star-Queen Nebula.'

Serpens
Hyginus. Poeticon
Astronomicon,
Venice, 1485.

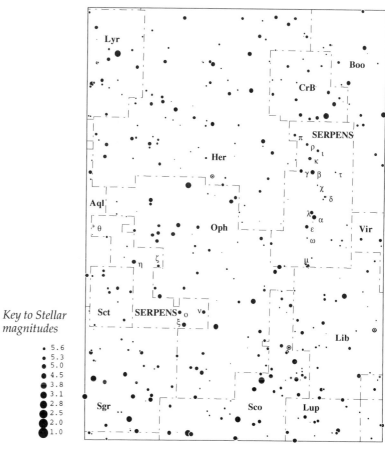

Key to Stellar
magnitudes

. 5.6
. 5.3
. 5.0
. 4.5
. 3.8
. 3.1
. 2.8
. 2.5
. 2.0
. 1.0

Sextans

Meaning:	The Sextant
Pronunciation:	sex' tans
Abbreviation:	Sex
Possessive form:	Sextantis (sex tan' tiss)
Asterisms:	none
Bordering constellations:	Crater, Hydra, Leo
Overall brightness:	1.595 (88)
Central point:	RA = 10h14m Dec. = –2°
Directional extremes:	N = +7° S = –11° E = 10h49m W = 9h39m
Messier objects:	none
Meteor showers:	Daytime Sextantids (29 Sep)
Midnight culmination date:	22 Feb
Bright stars:	none
Named stars:	none
Near stars:	Ross 446 (104), LFT 698 (132), LFT 709 (159), BD-3°2870 (165), LFT 729 (180)
Size:	313.51 square degrees (0.760% of the sky)
Rank in size:	47
Solar conjunction date:	26 Aug
Visibility:	completely visible from latitudes: +79° to –83° portions visible worldwide
Visible stars:	(number of stars brighter than magnitude 5.5): 5
Interesting facts:	(1) One of seven constellations still in use invented by Johannes Hevelius. In 1690, this group was included in a star atlas which accompanied his stellar catalog.

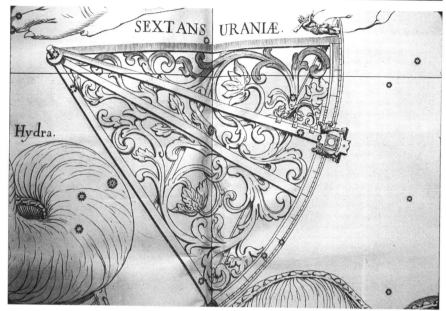

Sextans
Hevelius, Johannes.
Firmamentum
Sobiescianum, sive
Uranographia,
totum Coelum
Stellatum, *Gdansk,
1690. This is the first
appearance of this
constellation on any
star map.*

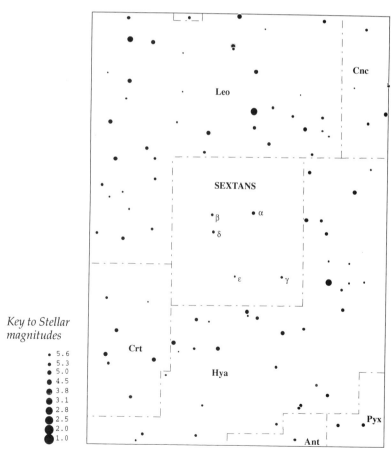

Key to Stellar
magnitudes

289

Taurus

Meaning:	The Bull
Pronunciation:	tor′ us
Abbreviation:	Tau
Possessive form:	Tauri (tor′ ee)
Asterisms:	The Heavenly G, The Hyades, The Pleiades, The V, The Winter Octagon, The Winter Oval

Bordering constellations: Aries, Auriga, Cetus, Eridanus, Gemini, Orion, Perseus

Overall brightness: 12.292 (12)

Central point: RA = 4h39m   Dec. = +15.5°

Directional extremes: N = +31°   S = 00°   E = 5h58m   W = 3h20m

Messier objects: M1, M45

Meteor showers: Daytime β Taurids (29 Jun)
S. Taurids (3 Nov)
N. Taurids (13 Nov)

Midnight culmination date: 30 Nov

Bright stars: α (14), β (27), η (140), ζ (167)

Named stars: Ain (ε), Alcyone (ζ), Aldebaran (α), Asterope (21), Atlas (27), Celaeno (16), Electra (17), El Nath (β), Hyadem I (γ), Hyadem II (δ¹), Maia (20), Merope (23), Nath (β), Pleione (28), Sterope (21), Taygeta (19)

Near stars: none

Size: 797.25 square degrees (1.933% of the sky)

Rank in size: 17

Solar conjunction date: 2 Jun

Visibility: completely visible from latitudes: N of −59°
portions visible worldwide

Visible stars: (number of stars brighter than magnitude 5.5): 98

Interesting facts:
(1) α Tau, or Aldebaran, is one of the four Royal Stars of the ancient Persians.

(2) β Tau was once a star 'shared' between the constellations of Taurus and Auriga. Pre-twentieth century star catalogs often list this star as γ Aur. Since the Belgian astronomer Eugène Delporte's *Délimitation Scientifique des Constellations* was adopted as the standard for constellation boundaries it has been assigned to Taurus.

(3) M45, also known as the 'Pleiades' or the 'Seven Sisters,' is the brightest galactic, or open, cluster in the sky. It is one of the few objects on Messier's list which does not possess a corresponding NGC number, presumably because it is so bright. Both Aratos and Eudoxus mention the Pleiades as a distinct constellation. (See list entitled 'The original 48 constellations.')

(4) Approximately 1° northwest of ζ Tau lies M1, the Crab Nebula. It received its name in the mid-nineteenth century when Lord Rosse noted that its extended filaments resembled the pincers of a crab. M1 is the gaseous remnant of a supernova which became visible in the year 1054. It is the brightest supernova remnant in the sky.

(5) It was within Taurus that the Italian astronomer Piazzi made the discovery of the first asteroid, Ceres, on New Year″s Day, 1801.

Taurus
Flamsteed, John. Atlas
Coelestis, *London,*
1729.

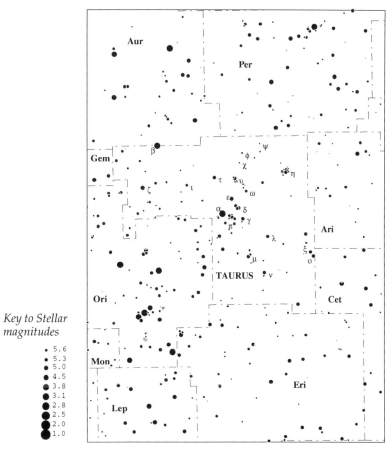

Key to Stellar
magnitudes

· 5.6
· 5.3
· 5.0
• 4.5
• 3.8
• 3.1
● 2.8
● 2.5
● 2.0
● 1.0

291

Telescopium

Meaning:	The Telescope
Pronunciation:	tel es koe' pee um
Abbreviation:	Tel
Possessive form:	Telescopii (tel es koe' pee ee)
Asterisms:	none
Bordering constellations:	Ara, Corona Australis, Indus, Pavo, Sagittarius
Overall brightness:	6.759 (49)
Central point:	RA = 19h16m Dec. = −51°
Directional extremes:	N = −45° S = −57° E = 20h26m W = 18h06m
Messier objects:	none
Meteor showers:	none
Midnight culmination date:	10 Jul
Bright stars:	none
Named stars:	none
Near stars:	none
Size:	251.51 square degrees (0.610% of the sky)
Rank in size:	57
Solar conjunction date:	10 Jan
Visibility:	completely visible from latitudes: S of +33°
	completely invisible from latitudes: N of +45°
Visible stars:	(number of stars brighter than magnitude 5.5): 17
Interesting facts:	(1) This was one of the 14 constellations invented by Lacaille during his stay at the Cape of Good Hope in 1751–2.

Telescopium (labeled 'le Telescope' on this map)
Lacaille, Nicolas Louis de. Planisphere contenant les Constellations Celestes, found in Mémoires Académie Royale des Sciences, *Paris, 1752 (published in 1756). This constellation was invented by Lacaille and the photo shows its first appearance on any star map.*

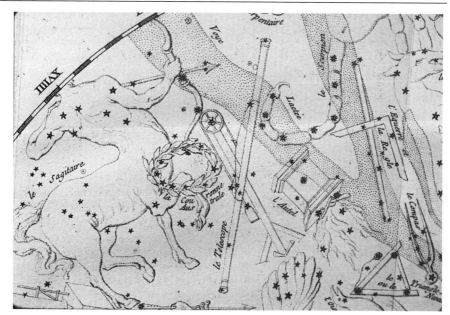

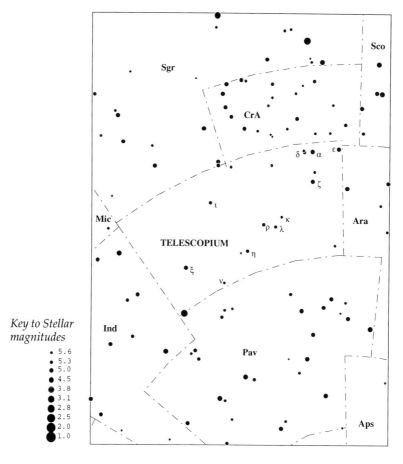

Key to Stellar magnitudes

293

Triangulum

Meaning:	The Triangle
Pronunciation:	try ang' yoo lum
Abbreviation:	Tri
Possessive form:	Trianguli (try ang' yoo lee)
Asterisms:	none

Bordering constellations: Andromeda, Aries, Perseus, Pisces

Overall brightness:	9.101 (27)
Central point:	RA = 2h08m Dec. = +31°

Directional extremes: N = +37°   S = +25°   E = 2h48m   W = 1h29m

Messier objects:	M33
Meteor showers:	none

Midnight culmination date: 23 Oct

Bright stars:	β (166)
Named stars:	Caput Trianguli (α), Mothallah (α)
Near stars:	none
Size:	131.85 square degrees (0.320% of the sky)
Rank in size:	78

Solar conjunction date: 24 Apr

Visibility:	completely visible from latitudes: N of –53°
	completely invisible from latitudes: S of –65°
Visible stars:	(number of stars brighter than magnitude 5.5): 12

Interesting facts: (1) Probably the hardest-to-see, bright deep-sky object is found within the boundaries of this constellation. This is M33, the Pinwheel Galaxy. It has a total light output equal to a star with magnitude 5.8, but even moderate instruments reveal scant detail. By far, the best views are through instruments using eyepieces which given low magnification. In extremely dark skies, this object can be glimpsed by some (including this writer) with the unaided eye, making it – and not M31 in Andromeda – the furthest object visible to the naked eye.

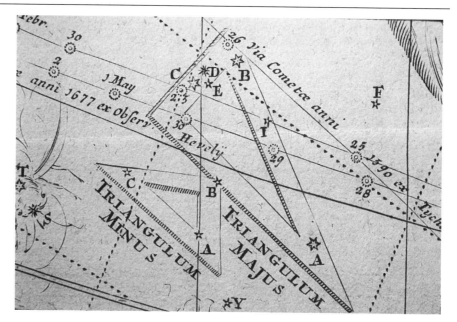

Triangulum Doppelmayr, Johann Gabriel. Atlas Coelestis, Nuremburg, 1742.

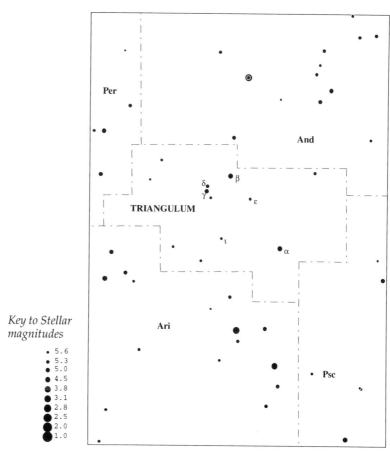

Key to Stellar magnitudes

295

Triangulum Australe

Meaning: The Southern Triangle

Pronunciation: try ang' yoo lum os trail'

Abbreviation: TrA

Possessive form: Trianguli Australis (try ang' yoo lee os tral' iss)

Asterisms: The Three Patriarchs

Bordering constellations: Apus, Ara, Circinus, Norma

Overall brightness: 10.911 (14)

Central point: RA = 15h59m   Dec. = −65°

Directional extremes: N = −60°   S = −70°   E = 17h09m   W = 14h50m

Messier objects: none

Meteor showers: none

Midnight culmination date: 23 May

Bright stars: α (42), β (136), γ (145)

Named stars: Atria (α)

Near stars: none

Size: 109.98 square degrees (0.267% of the sky)

Rank in size: 83

Solar conjunction date: 21 Nov

Visibility: completely visible from latitudes: S of +20°
completely invisible from latitudes: N of +30°

Visible stars: (number of stars brighter than magnitude 5.5): 12

Interesting facts: (1) This constellation was first described by the Italian navigator Amerigo Vespucci in 1503. Triangulum Australe was 'reintroduced' by Keyser and de Houtman during their travels of 1595–7, but they did not invent it.

Triangulum Australe
Bode, Johann Elert.
Uranographia Sive
Astrorum
Descriptio, *Berlin,*
1801.

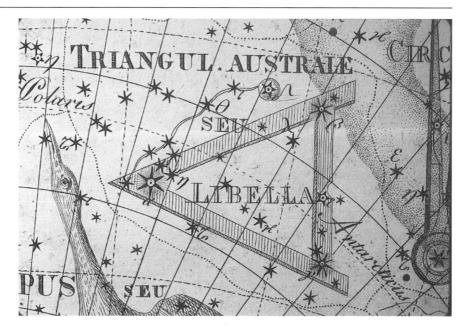

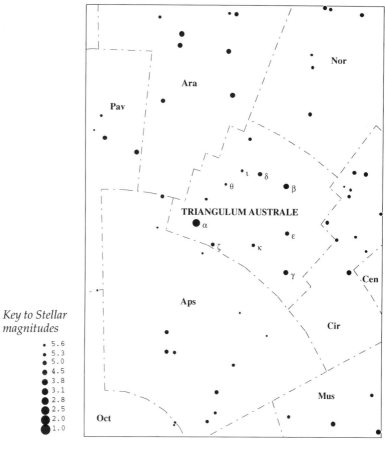

Key to Stellar
magnitudes

- 5.6
- 5.3
- 5.0
- 4.5
- 3.8
- 3.1
- 2.8
- 2.5
- 2.0
- 1.0

297

Tucana

Meaning:	The Toucan
Pronunciation:	too kan' uh
Abbreviation:	Tuc
Possessive form:	Tucanae (too kan' eye)
Asterisms:	none
Bordering constellations:	Grus, Hydrus, Indus, Octans, Phoenix
Overall brightness:	5.092 (74)
Central point:	RA = 23h43m Dec. = −66.5°
Directional extremes:	N = −57° S = −76° E = 1h22m W = 22h05m
Messier objects:	none
Meteor showers:	none
Midnight culmination date:	17 Sep
Bright stars:	α (139)
Named stars:	none
Near stars:	ζ Tuc (89), LFT 117 (142)
Size:	294.56 square degrees (0.714% of the sky)
Rank in size:	48
Solar conjunction date:	18 Mar
Visibility:	completely visible from latitudes: S of +14°
	completely invisible from latitudes: N of +33°
Visible stars:	(number of stars brighter than magnitude 5.5): 15

Interesting facts: (1) This is one of 11 constellations invented by Pieter Dirksz Keyser and Frederick de Houtman, during the years 1595–7.

(2) The most magnificent object within the confines of Tucana is the Small Magellanic Cloud (SMC), also known as 'Nubecula Minor.' This is the smaller, and more distant, satellite galaxy of the Milky Way. Its mass is 'only' about 2000 million times the mass of the Sun, and it lies at a distance of approximately 200 000 light years. It is interesting that this object, despite its brightness, was given an NGC number. The Large Magellanic Cloud does not possess such a designation; nor does M45, the Pleiades. The SMC has a total, integrated visual magnitude of 1.5.

(3) About 1° from the western edge of the SMC lies the second finest globular cluster in the entire sky, surpassed in appearance only by the ω Cen cluster. It is known as 47 Tuc. This object, with a brightness equal to that of a star of magnitude 4.5, was designated as a stellar object on early maps, just like ω Cen. Its position is near the SMC, but it lies within the outer confines of the Milky Way, and thus is considered part of our galaxy.

(4) On 28 December 1969, Comet Bennett, one of the great comets of the twentieth century was discovered in Tucana. During March, 1970, the comet reached its maximum brightness of approximately zero magnitude.

Tucana
Bayer, Johann.
Uranometria,
Augsburg, 1603. This
constellation was
invented by de
Houtman and Keyser
in 1596. It was first
illustrated on a globe
by Plancius, which has
not survived. This
photo from Bayer's
map, therefore, shows
the earliest existing
picture of this
constellation.

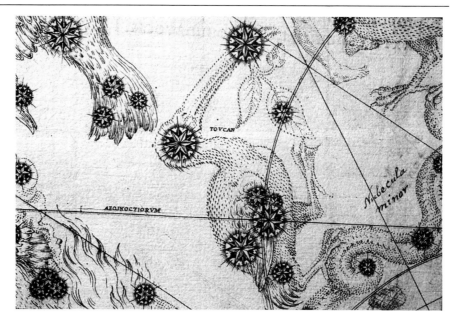

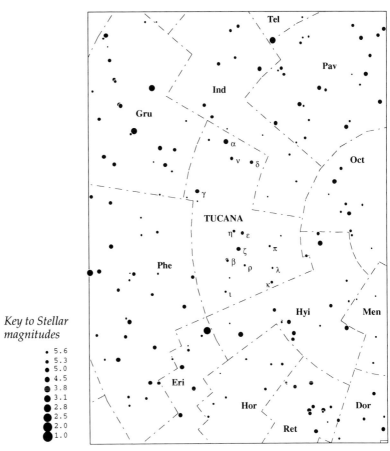

Key to Stellar
magnitudes

· 5.6
· 5.3
· 5.0
· 4.5
● 3.8
● 3.1
● 2.8
● 2.5
● 2.0
● 1.0

Ursa Major

Meaning:	The Great Bear
Pronunciation:	er' suh may' jor
Abbreviation:	UMa
Possessive form:	Ursae Majoris (er' sigh muh jor' iss)
Asterisms:	The Arc, The Bier, The Big Dipper, The Horse and Rider, The Pointers
Bordering constellations:	Boötes, Camelopardalis, Canes Venatici, Coma Berenices, Draco, Leo, Leo Minor, Lynx
Overall brightness:	5.548 (69)
Central point:	RA = 11h16m Dec. = +51°
Directional extremes:	N = +73° S = +29° E = 14h27m W = 8h05m
Messier objects:	M40, M81, M82, M97, M101, M108, M109
Meteor showers:	Ursids (22 Dec)
Midnight culmination date:	11 Mar
Bright stars:	ε (32), α (34), η (38), ζ¹ (71), β (77), γ (84), ψ (171), μ (178), ι (196), h (200)
Named stars:	Alcor (80), Alioth (ε), Alkaid (η), Alula Australis (ξ), Alula Borealis (ν), Benetnash (η), Dnoces (ι), Dubhe (α), Kaffa (δ), Megrez (δ), Merak (β), Mizar (ζ), Muscida (π²), Phad (γ), Phecda (γ), Talitha (ι), Tania Australis (μ), Tania Boraelis (λ)
Near stars:	Lalande 21185 (5), BD+50°1725 (29), WX UMa A-B (44), LFT 634–635 (67), ξ UMa A-B (110), SZ UMa (126), 61 UMa (152), Groombridge 1830 (153)
Size:	1279.66 square degrees (3.102% of the sky)
Rank in size:	3
Solar conjunction date:	10 Sep
Visibility:	completely visible from latitudes: N of –17° completely invisible from latitudes: S of –61°
Visible stars:	(number of stars brighter than magnitude 5.5): 71

Interesting facts: (1) ζ UMa, named 'Mizar,' along with 80 UMa, called 'Alcor,' form a relatively close visual double star in the bend of the handle of the Big Dipper. Mizar itself was the first double star to be telescopically discovered. This occurred in 1650. In 1889, the primary of the Mizar pair became the first spectroscopically discovered binary star. Since then, Alcor has also been classified as a spectroscopic binary. Thus, no less than five stars comprise this magnificent stellar system.

(2) For observers with moderate-sized instruments, two of the brightest galaxies in the sky may be compared and contrasted in the same field. These are M81 and M82. They lie approximately 2° to the east of the star 24 UMa. M81 is a classic spiral galaxy and M82 is an irregular. However, this object is unusual even within the irregular classification. M82 *appears* to have undergone a titanic explosion which has rent this galaxy asunder. Many studies, however, seem to indicate that this is merely a galaxy where a great deal of cosmic activity is occurring. In fact, some have labeled M82 a 'starburst galaxy,' a galaxy in which a tremendous amount of star formation is taking place.

Ursa Major
Cellarius, Andreas.
Harmonia
Macrocosmica sev
Atlas Universalis et
Novus, *Amsterdam,*
1661.

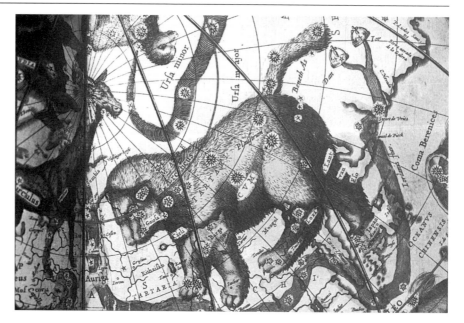

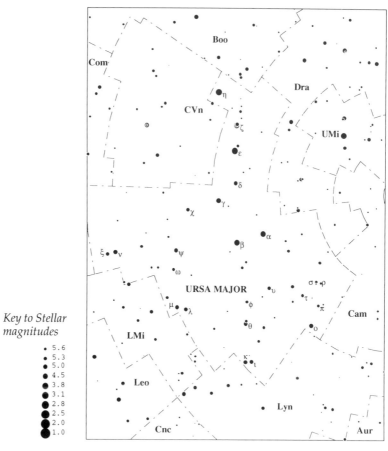

Key to Stellar
magnitudes

301

Ursa Minor

Meaning:	The Bear Cub
Pronunciation:	er' sa my' nor
Abbreviation:	UMi
Possessive form:	Ursae Minoris (er' sigh muh nor' iss)
Asterisms:	The Guardians of the Pole, The Little Dipper

Bordering constellations: Camelopardalis, Cepheus, Draco

Overall brightness: 7.035 (45)

Central point: RA = (circumpolar)   Dec. = +77.5°

Directional extremes: N = +90° S = +65° E = circumpolar W = circumpolar

Messier objects: none

Meteor showers: none

Midnight culmination date: 13 May

Bright stars: α (48), β (55), γ (180)

Named stars: Cynosaura (α), Kochab (β), Pherkad (γ), Pherkard (δ), Polaris (α), Yildun (δ)

Near stars: none

Size: 255.86 square degrees (0.620% of the sky)

Rank in size: 56

Solar conjunction date: 21 Sep

Visibility: completely visible from latitudes: N of +00°
completely invisible from latitudes: S of –25°

Visible stars: (number of stars brighter than magnitude 5.5): 18

Interesting facts: (1) α UMi, or Polaris, is undoubtedly the most famous single star in the sky. Often lecturers and writers go to great lengths to point out that Polaris is not the brightest star in the sky, in fact, it ranks 'only' 48th. This ranking may seem low, but it does place Polaris in the top 2% of all visible stars. At the time of writing, Polaris is 0.77° from the north celestial pole. Because of the Earth's motion of precession, Polaris is moving closer to this point. It will be at its nearest in the year AD 2102, when it will lie at a distance of 0.46°.

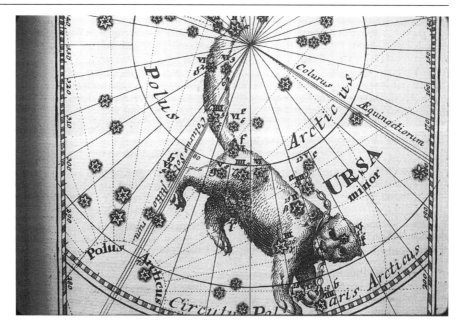

Ursa Minor
Thomas, Corbinianus.
Mercurii
Philosophici
Firmanentum
Firmianum,
Frankfurt and Leipzig,
1730.

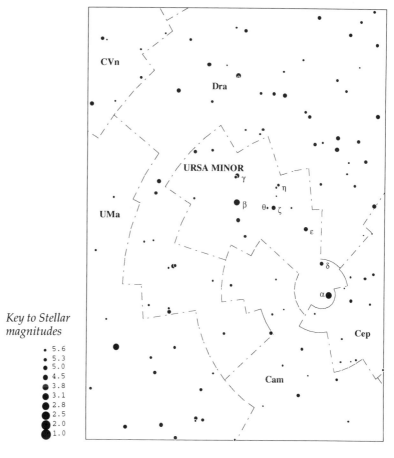

Key to Stellar
magnitudes

- 5.6
- 5.3
- 5.0
- 4.5
- 3.8
- 3.1
- 2.8
- 2.5
- 2.0
- 1.0

Vela

Meaning:	The Sail (of Argo Navis)
Pronunciation:	vay' luh
Abbreviation:	Vel
Possessive form:	Velorum (vee lor' um)
Asterisms:	none

Bordering constellations: Antlia, Carina, Centaurus, Puppis, Pyxis

Overall brightness:	15.211 (4)
Central point:	RA = 9h43m Dec. = –47°
Directional extremes:	N = –37° S = –57° E = 11h24m W = 8h02m
Messier objects:	none
Meteor showers:	none

Midnight culmination date: 13 Feb

Bright stars:	λ (62), κ (90), μ (109), N (192)
Named stars:	Alsuhail (λ), Markeb (χ), Regor (γ)
Near stars:	LFT 682 (158)
Size:	499.65 square degrees (1.211% of the sky)
Rank in size:	32

Solar conjunction date: 18 Aug

Visibility:	completely visible from latitudes: S of +33°
	completely invisible from latitudes: N of +53°
Visible stars:	(number of stars brighter than magnitude 5.5): 76

Interesting facts: (1) One of three constellations into which Lacaille divided the ancient constellation of Argo Navis. The other two 'sub-constellations' are Carina and Puppis.

(2) A very unusual object is located within this constellation. It is known as the 'Vela Pulsar.' It was the second such object to be observed optically, although radio telescopes had discovered it a number of years before. It was seen in 1977, ten years after the optical discovery of the pulsar within the Crab Nebula. The Vela Pulsar is generally considered the faintest luminous object which has yet been observed. It has a visual magnitude of 26, and an absolute magnitude estimated to be only 34.5.

(3) In visually scanning the Milky Way within Vela, we observe the only complete break in this diffuse band of light which marks the plane of our galaxy. Known as the 'Great Rift,' this area is due to the presence of dark nebulae composed of dust and cold gas.

Vela (labeled 'la Voilure' on this map) Vaugondy, Robert de. **Hémisphère Céleste Antarctique...**, *Paris, 1764. This constellation was invented by Lacaille and included in his star catalog, but not pictured on any of his star charts. This photo from Robert de Vaugondy's map, therefore, shows the first appearance of Vela as a separate constellation.*

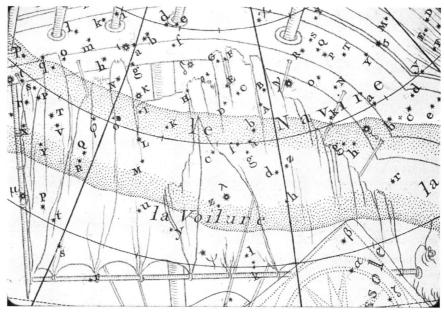

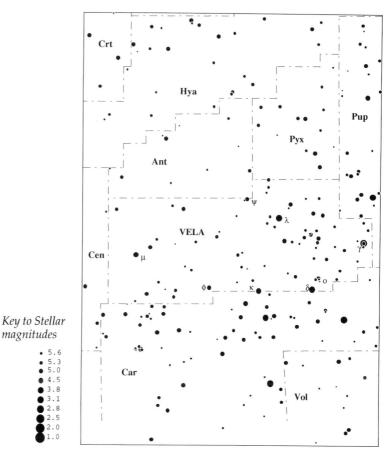

Key to Stellar magnitudes

- 5.6
- 5.3
- 5.0
- 4.5
- 3.8
- 3.1
- 2.8
- 2.5
- 2.0
- 1.0

Virgo

Meaning:	The Virgin
Pronunciation:	ver' go
Abbreviation:	Vir
Possessive form:	Virginis (ver' jin iss)
Asterisms:	The Diamond, The Spring Triangle, The Y

Bordering constellations: Boötes, Coma Berenices, Corvus, Crater, Hydra, Leo, Libra, Serpens

Overall brightness:	4.481 (79)
Central point:	RA = 13h21m Dec. = –4°

Directional extremes: N = +14°   S = –22°   E = 15h08m   W = 11h35m

Messier objects:	M49, M58, M59, M60, M61, M84, M86, M87, M89, M90, M104
Meteor showers:	Virginids (26 Mar)
	α Virginids (9 Apr)
	μ Virginids (25 Apr)

Midnight culmination date: 11 Apr

Bright stars:	α (16), γ (117), ε (133)
Named stars:	Alaraph (β), Arich (γ), Auva (δ), Azimech (α), Heze (ζ), Minelauva (δ), Porrima (γ), Spica (α), Syrma (ι), Vindemiatrix (ε), Zania (η), Zavijava (β)
Near stars:	Ross 128 (12), Wolf 424 A-B (27), Wolf 489 (94), Ross 490 (101), 61 Vir (127), Wolf 437 (149), Wolf 457 (162), β Vir (195), γ Vir A-B (198)
Size:	1294.43 square degrees (3.138% of the sky)
Rank in size:	2

Solar conjunction date: 12 Oct

Visibility:	completely visible from latitudes: +68° to –76°
	portions visible worldwide
Visible stars:	(number of stars brighter than magnitude 5.5): 58

Interesting facts: (1) The autumnal equinox, that point on the celestial sphere where the Sun crosses the celestial equator heading south, lies in Virgo. Its position is approximately 3° southeast of β Vir.

(2) The best known cluster of galaxies, called the 'Virgo Cluster,' lies in the region of sky just to the east of ε Vir, near the border of Coma Berenices. This area has been labeled 'The Realm of the Nebulae,' a reference to times when the nature of the galaxies was not well understood. The Virgo Cluster lies at a distance of 50–60 million light years and contains about 2500 members, most of which are spiral galaxies. The most famous members of this cluster are M104, the so-called 'Sombrero Galaxy,' and M87, which seems to be ejecting a jet of material from its nucleus.

(3) One of the most famous double stars in the sky is γ Vir, also known as Porrima. This system is so celebrated that no less an observer than Smyth wrote a poem containing 22 stanzas entitled, 'A Farewell to the Double Star γ Virginis at the Epoch of 1858.' Since then many a college student, including this writer, has taken micrometric observations of this pair to calculate its orbital elements.

(4) Slightly less than 5° to the northwest of γ Vir lies 3C273, the brightest quasar in the entire sky, shining with an apparent magnitude of 12.8. This is also the first quasar whose spectral lines were deciphered (by Maarten Schmidt) to show an incredible redshift. The approximate distance to 3C273 is 3000 million light years.

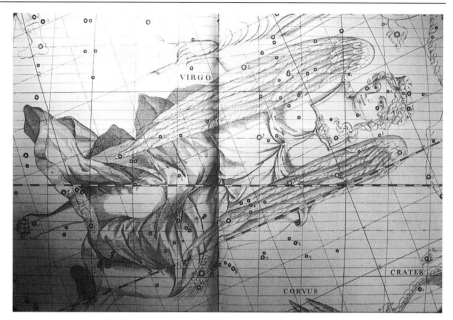

Virgo
Flamsteed, John. Atlas
Coelestis, *London,*
1729.

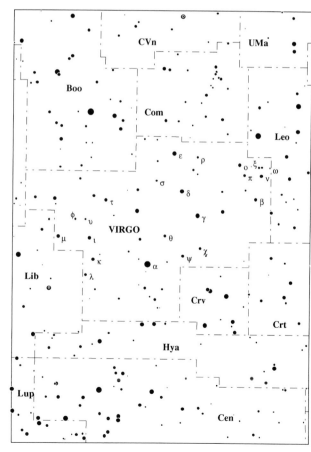

Key to Stellar
magnitudes

- 5.6
- 5.3
- 5.0
- 4.5
- 3.8
- 3.1
- 2.8
- 2.5
- 2.0
- 1.0

Volans

Meaning:	The Flying Fish
Pronunciation:	voe' lans
Abbreviation:	Vol
Possessive form:	Volantis (voe lan' tiss)
Asterisms:	none

Bordering constellations: Carina, Chamaeleon, Dorado, Mensa, Pictor

Overall brightness:	9.904 (19)
Central point:	R.A. = 7h48m Dec. = −69.5°
Directional extremes:	N = −64° S = −75° E = 9h02m W = 6h35m
Messier objects:	none
Meteor showers:	none

Midnight culmination date: 18 Jan

Bright stars:	none
Named stars:	none
Near stars:	L 97-12 (54)
Size:	141.35 square degrees (0.343% of the sky)
Rank in size:	76

Solar conjunction date: 19 Jul

Visibility:	completely visible from latitudes: S of +15°
	completely invisible from latitudes: N of +26°
Visible stars:	(number of stars brighter than magnitude 5.5): 14
Interesting facts:	(1) This is one of 11 constellations invented by Pieter Dirksz Keyser and Frederick de Houtman, during the years 1595–7.

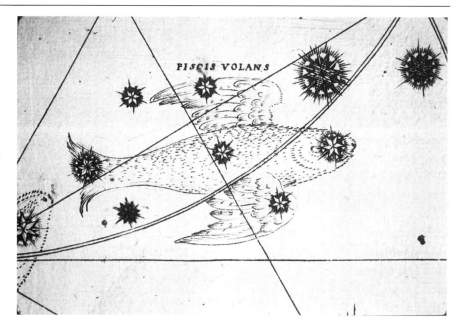

Volans (labeled 'Piscis Volans' on this map) Bayer, Johann. Uranometria, *Augsburg, 1603. This constellation was invented by de Houtman and Keyser in 1596. It was first illustrated on a globe by Plancius, which has not survived. This photo from Bayer's map, therefore, shows the earliest existing picture of this constellation.*

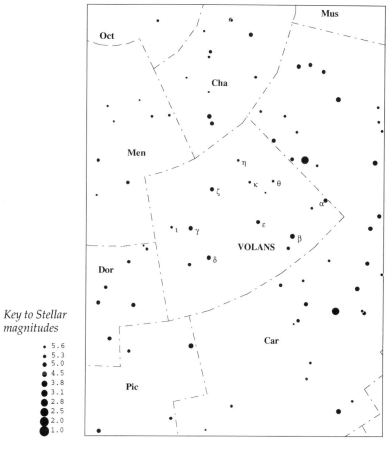

Key to Stellar magnitudes

Vulpecula

Meaning:	The Fox
Pronunciation:	vul pek' yoo luh
Abbreviation:	Vul
Possessive form:	Vulpeculae (vul pek' yoo lye)
Asterisms:	none
Bordering constellations:	Cygnus, Delphinus, Hercules, Lyra, Pegasus, Sagitta
Overall brightness:	10.814 (15)
Central point:	RA = 20h12m Dec. = +24°
Directional extremes:	N = +29° S = +19° E = 21h28m W = 18h56m
Messier objects:	M27
Meteor showers:	none
Midnight culmination date:	25 Jul
Bright stars:	none
Named stars:	none
Near stars:	Ross 165 A-B (191)
Size:	268.17 square degrees (0.650% of the sky)
Rank in size:	55
Solar conjunction date:	24 Jan
Visibility:	completely visible from latitudes: N of –61°
	completely invisible from latitudes: S of –71°
Visible stars:	(number of stars brighter than magnitude 5.5): 29

Interesting facts:
(1) One of seven constellations still in use invented by Johannes Hevelius. In 1690, this group was included in a star atlas which accompanied his stellar catalog.

(2) The first pulsar was discovered within the boundaries of this constellation. Designated PSR 1919+21, it was found in 1967 by Jocelyn Bell while she was a graduate student at Cambridge University. Because of the regular pulsations of this object, which occur every one and one-third seconds, some astronomers were convinced that this represented a signal from extra-terrestrial intelligence. This pulsar was, therefore, given a different designation: 'LGM.' This was an acronym for three words which represented the hopes (or fears) of many scientists – 'Little Green Men.'

(3) M27, the deep-sky object generally regarded to be the finest planetary nebula in the sky, lies in Vulpecula. With a visual magnitude of 7.6, M27 has been dubbed 'The Dumbbell' due to its characteristic shape. It is quite an easy object to locate and a wonderful sight in instruments of all sizes.

Vulpecula
Hevelius, Johannes.
Firmamentum
Sobiescianum, sive
Uranographia,
totum Coelum
Stellatum, *Gdansk,*
1690. This is the first
appearance of this
constellation on any
star map.

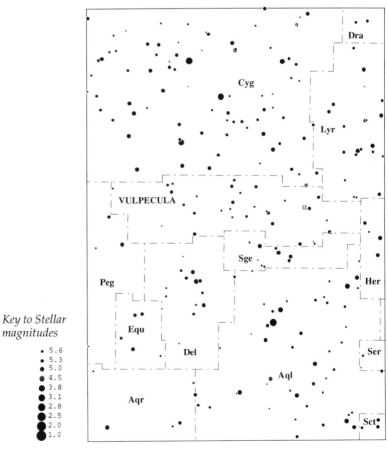

Key to Stellar
magnitudes

311

Glossary

absolute magnitude	the apparent magnitude of a celestial object if the object were at a distance of 10 parsecs; the true brightness of a celestial object; luminosity.
achronical rising	the rising of a star or constellation at (or just following) sunset. When the rising is achronical the setting is cosmic.
achronical setting	the setting of a star or constellation at (or just following) sunset. When the setting is achronical the rising is cosmic.
achronychian rising	the rising of a star or constellation in evening twilight; the last observable rising of a star or constellation (in the annual cycle). After this point, the object is above the horizon when it first becomes visible. *See also* achronical rising)
achronychian setting	the setting of a star or constellation in morning twilight; the last observable setting of a star or constellation (in the annual cycle). Before this point, the object may be seen to set before morning twilight obscures viewing. *See also* achronical setting)
Almagest	a 13 volume astronomical work compiled by Ptolemy of Alexandria around AD 140, which included a star catalog giving the positions and approximate magnitudes of 1022 stars. The word 'Almagest' comes from the Arabic and means 'the greatest'. The Greek name for this work is the 'Syntaxis.'
al-Sufi (903–986)	A Persian astronomer who compiled a catalog of approximately 1,000 stars. The stars were designated by position, color, and brightness.
altitude	the angular distance of a point or celestial object above or below the horizon. It is measured along the vertical circle through the body from 0° (on the horizon) to 90° (at the zenith). Negative values correspond to objects which lie below the horizon.
angular diameter	the observed diameter of a celestial body, expressed in degrees, minutes, and/or seconds of arc.
angular distance	the observed distance between two celestial bodies expressed in degrees, minutes, and/or seconds of arc.
antapex	see solar antapex.
apastron	the point in any orbit around a star that is farthest from the star.
apex	*see* solar apex.
apparition	the period of time during which a celestial body may be observed.
apparent magnitude	the brightness of a celestial object as seen from earth, irrespective of its true brightness.

Argelander, Friedrich Wilhelm August (1799–1875)	a Prussian astronomer who did pioneering work on the magnitude system of stellar brightnesses. He also compiled a star catalog of more than 300 000 northern stars known as the *Bonner Durchmusterung*.
asterism	an unofficial, recognizable grouping of visible stars. The stars within an asterism may belong to one constellation (e.g. the 'Big Dipper,' within the constellation Ursa Major) or several constellations (e.g. the 'Summer Triangle,' composed of stars from the constellations Lyra, Cygnus, and Aquila).
astronomical horizon	the great circle on the celestial sphere 90° from the zenith.
astronomical twilight	the condition of solar illumination at a point on the Earth's surface before sunrise or after sunset when the Sun's zenith distance is 108°.
atmospheric refraction	the bending of light passing obliquely through a body's atmosphere; in the case of the earth, the result is that celestial bodies appear to be displaced towards the zenith, with the amount of displacement increasing with the object's zenith distance.
autumn (or, autumnal) equinox	*see* 'September equinox.'
azimuth	the angular distance to an object measured eastwards along the horizon from the north to the intersection of the object's vertical circle; varies from 0° to 360°. Thus, an object due east would have an azimuth of 90°, and an object due west would have an azimuth of 270°.
Bayer, Johannes (1564–1617)	a Bavarian astronomer who first designated the stars within constellations by Greek letters, in approximate decreasing order of brightness.
Beg, Ulugh (1394–1449)	a Persian astronomer who made detailed observations of the Moon and planets, determined the inclination of the ecliptic, and compiled a very precise catalog of 1012 stars.
Bessel, Friedrich Wilhelm (1784–1846)	A German astronomer who first observed stellar parallax – in 1838, he measured the distance to the star 61 Cygni.
celestial equator	the intersection of the equatorial plane of the Earth with the celestial sphere; the projection of the Earth's equator onto the sky.
celestial sphere	the apparent background of the stars, assumed to be of infinite extent in all directions; the sky.
civil twilight	the condition of solar illumination at a point on the Earth's surface before sunrise or after sunset when the Sun's zenith distance is 96°.
conjunction	the alignment of two celestial objects such that the difference in their longitude, as seen from Earth, is 0°. Two objects may also be in conjunction in right ascension. When one of the objects is the Sun, 'conjunction' denotes when the other object is in line with the Sun, and therefore invisible.
constellation	one of 88 officially recognized areas of the sky according to the International

Astronomical Union (1928); also may refer to one of the 88 officially recognized groupings of visible stars within those areas.

cosmic rising the rising of a star at (or just following) sunrise. When the rising is cosmic the setting is achronical.

cosmic setting the setting of a star at (or just following) sunrise. When the setting is cosmic the rising is achronical.

culmination the passage of a celestial body across an observer's meridian. 'Upper culmination' (also called 'transit') is the crossing nearer to the observer's zenith; 'Lower culmination' is the crossing further from the observer's zenith.

December solstice that instant when the Sun achieves minimum declination; the point on the ecliptic where the Sun's declination is at a minimum, having celestial coordinates RA =18h Dec. = −23.5° approximately. (Note this point is the "winter" solstice only in the northern hemisphere of Earth.)

declination (Dec.) an Earth-centered angle measured perpendicularly from the celestial equator to a point on the celestial sphere. Declination is positive if the object or point is north of the celestial equator and negative if the object or point is south of the celestial equator.

ecliptic the great circle described by the Sun's annual path on the celestial sphere; the mean plane of the Earth's orbit around the Sun.

equinoctial colure the great circle passing through both celestial poles and intersecting the ecliptic at both equinoxes.

equinox either of two points on the ecliptic, lying at right ascension 0 hours (March equinox) and 12 hours (September equinox).

Flamsteed, John (1646–1719) The first Astronomer Royal of England, his most famous work was a catalog of approximately 3000 stars with very precise right ascension coordinates.

Gegenschein (or, counterglow) a very faint glow of light visible at the position of the ecliptic 180° from the Sun. It is believed to be caused by sunlight reflected from tiny interplanetary particles.

heliacal rising the rising of a star or planet on the date when it can first be observed in the morning sky during the twilight before sunrise.

heliacal setting the setting of a star or planet on the date when it can last be observed in the evening sky during the twilight after sunset.

Hevelius, Johannes (1611–1687) a German astronomer who published the first detailed map of the Moon and a celestial atlas in which he introduced the constellations of Canes Venatici, Lacerta, Leo Minor, Lynx, Scutum, Sextans, and Vulpecula.

Hipparchus (c.146 – c.127 BC) a Greek astronomer who compiled the first star catalog. Hipparchus' catalog contained 850 stars arranged by brightness, and it was this arrangement which forms the basis of the present magnitude system of stellar brightnesses.

horizon where the celestial sphere intersects the Earth at every point; where the sky meets the Earth . *See also* astronomical horizon."

hour circle any of an infinite number of circles on the celestial sphere which intersect both the north and south celestial poles.

de Houtman, Frederick (1540–1627) Dutch navigator responsible, with Pieter Dirksz Keyser, for the formation of the constellations of Apus, Chamaeleon, Dorado, Grus, Hydrus, Indus, Musca, Pavo, Phoenix, Tucana, and Volans.

June solstice that instant when the sun achieves maximum declination; the point on the ecliptic where the Sun's declination is at a maximum, having celestial coordinates of RA = 6h Dec. = +23.5°, approximately. (Note this point is the 'summer' solstice only in the northern hemisphere of Earth.)

Keyser, Pieter Dirksz (d. 1596) Dutch navigator responsible, with Frederick de Houtman, for the formation of the constellations of Apus, Chamaeleon, Dorado, Grus, Hydrus, Indus, Musca, Pavo, Phoenix, Tucana, and Volans.

Lacaille, Nicolas Louis de (1713–1762) A French astronomer who compiled a detailed catalog of stars of the southern hemisphere and introduced the constellations Antlia, Caelum, Carina, Circinus, Fornax, Horologium, Mensa, Microscopium, Norma, Octans, Pictor, Puppis, Pyxis, Reticulum, Sculptor, Telescopium, and Vela.

light year the distance that light, moving at approximately 186 000 miles per second, (300 000 km/s) travels in one year.

lower culmination *see* culmination.

magnitude a measure of the amount of light flux (or other radiation) received from a luminous celestial object. (*See also* absolute magnitude and apparent magnitude.)

March equinox that instant when the Sun, moving northerly, crosses the equatorial plane of the Earth; one of two points where the ecliptic and celestial equator meet, having celestial coordinates RA = 0h Dec. = 0°. (Note this point is the 'spring' equinox only in the northern hemisphere of Earth.

Mercator, Gerard (1512–1594) responsible for the first appearance of the constellation Coma Berenices, which appeared on a globe of his creation in 1551.

meridian the great circle passing through the observer's zenith and the celestial poles.

Messier, Charles (1730–1817) a French astronomer and comet hunter. His most famous work, a list of objects which was published to avoid confusion between those objects and possible comets, is still in use today.

Messier object one of a list of non-stellar objects compiled by Charles Messier, and added to by others. The present list contains 110 entries, each of which is designated by an "M" number (e.g., M11).

nadir the point on the celestial sphere that lies directly beneath the observer, on the observer's meridian; the point opposite the "zenith."

nautical twilight the condition of solar illumination at a point on the Earth's surface before sunrise or after sunset when the Sun's zenith distance is 102°.

nebula (Latin for 'cloud') a cloud of interstellar gas or dust.

NGC number a number given to any object (galaxy, star cluster, or nebula) found in J.L.E. Dreyer's *A New General Catalogue of Nebulae and Clusters of Stars, being the Catalogue of the late Sir John F.W. Herschel, Bart., revised, corrected, and enlarged*. This work was published as *Memoirs of the Royal Astronomical Society*, vol. xlix, part 1, London, 1888. Two supplements, also Memoirs of the RAS, were published by Dreyer. These were entitled *Index Catalog of Nebulae found in the Years 1888 to 1894, etc.* (London, 1895) and *Second Index Catalog of Nebulae and Clusters, etc.* (London, 1908). These three catalogs are now published as one work, known as the *Revised New General Catalog of Nonstellar Astronomical Objects*.

nightglow (airglow) a faint glow in the Earth's upper atmosphere caused by light emitted during the recombination of atoms and molecules following collisions with high energy particles and photons, mainly from the Sun. Termed nightglow when seen at night, and airglow at other times.

opposition the position of two celestial objects when their longitude (as seen from Earth) differs by 180°. When one of the objects is the Sun opposition means the other object is opposite the Sun in the sky, therefore visible all night long.

parallax *See* stellar parallax.

paranatellon (or, parantellon)a constellation that rises at the same time as another constellation in a particular location. e.g., Aquila is a paranatellon of Capricornus in certain locations, as both rise at the same time.

parsec the distance at which a star would have an annual parallax of one arc-second; at this distance the semimajor axis of the Earth's orbit (one astronomical unit) subtends an angle of one arc-second. One parsec = 30.857×10^{12} km = 206 265 astronomical units = 3.2616 light years.

perihelion the position of an object in solar orbit when it is closest to the Sun; the instant in a given orbit of a planet (or other body) when it is closest to the Sun.

precession the sweeping out of a cone by the spin axis of a rotating body when acted upon by a torque perpendicular to its spin axis. The Sun and Moon (and to a much lesser extent, the planets) attract the equatorial bulge of the nonspherical Earth causing the poles to precess about a line through the Earth's center perpendicular to the ecliptic plane; thus, the celestial poles describe circles approximately 23.5 degrees in radius on the celestial sphere.

proper motion the apparent angular motion per year of a star or other celestial object in a direction perpendicular to the line of sight.

Ptolemy, Claudius (73 AD–151 AD) a Greek astronomer and philosopher whose greatest work, the Almagest influenced astronomical thought for 15 centuries after its publication in the second century AD. His listing of 48 constellations in his star catalog formed the basis of celestial cartography. Only two of his constellations (Argo Navis and The Pleiades) are no longer in use today.

radial velocity the velocity of a star or other celestial object along the line of sight of the observer.

right ascension (RA) a geocentric spherical coordinate that is an angle measured eastwards along the celestial equator from the vernal equinox to the intersection of the hour circle passing through the body; usually expressed in hours, minutes, and seconds from 0 hours to 24 hours, where one hour of right ascension equals 15 degrees.

September equinox that instant when the Sun, moving southerly, crosses the equatorial plane of the earth; one of two points where the ecliptic and celestial equator meet, having celestial coordinates RA = 12h Dec. = 0°. (Note this point is the 'autumn' equinox only in the northern hemisphere of Earth.)

sidereal time 'star time' defined as the local hour angle of the vernal equinox; also defined as the right ascension of a (real or hypothetical) star on the local meridian.

solar antapex the point on the celestial sphere away from which the Sun and solar system are moving, relative to the stars in our vicinity. This point lies within the boundaries of the constellation of Columba, and has the approximate coordinates RA = 6h Dec. = –30°.

solar apex the point on the celestial sphere towards which the Sun and the solar system are moving, relative to the stars in our vicinity. This point lies within the boundaries of the constellation of Hercules, and has the approximate coordinates RA = 18h Dec. = +30°.

solstice either of two points on the ecliptic, lying at right ascension 6 hours (June solstice) and 18 hours (December solstice).

solstitial colure the great circle passing through both celestial poles and intersecting the ecliptic at both solstices.

spring equinox *see* March equinox.

star a self-luminous sphere of gas that generates energy by means of nuclear fusion reactions in its core.

stellar parallax the apparent angular displacement of a star or other celestial object that results from the revolution of the Earth about the Sun; numerically, this is the angle subtended by one astronomical unit at the distance of the particular object.

summer solstice *see* June solstice.

upper culmination *see* culmination.

vernal equinox that instant when the Sun, moving northerly, crosses the equatorial plane of the Earth; one of two points where the ecliptic and celestial equator meet, having celestial coordinates RA = 0h Dec. = 0°. (Note this point is also known as the March equinox and is the spring equinox only in the northern hemisphere of Earth.)

vertical circle any of an infinite number of circles on the celestial sphere which pass through the observer's zenith and intersect the horizon at a right angle.

visual magnitude *see* apparent magnitude.

winter solstice *see* December solstice.

zenith the point on the celestial sphere which lies 90° from all points on the horizon; the overhead point, lying on the observer's meridian; the point opposite the 'nadir'.

zenith distance the angle which a celestial body makes with the overhead point, as seen by an earthbound observer. An object on the observer's horizon has a zenith distance of 90°.

zodiac a band around the celestial sphere 18° in width and centered on the ecliptic.

References and sources

Astronomical Dictionary in Six Languages, Josip Kleczek, 1961, Academic Press, New York and London

Astronomy with an Opera Glass, Garrett P. Serviss, First edition, 1888, D. Appleton & Co., New York

The Biographical Dictionary of Scientists: Astronomers, David Abbott, general editor, First edition, 1984, Peter Bedrick Books, New York

The Bright Star Catalogue, Dorritt Hoffleit, Fourth revised edition, 1982, Yale University Observatory, New Haven, Connecticut

Burnham's Celestial Handbook - 3 Volumes, Robert Burnham, Jr., Revised edition, 1978, Dover Publications, New York

The Cambridge Deep-Sky Album, Jack Newton and Philip Teece, First edition, 1983, Cambridge University Press

A Catalogue of 1849 Stars with Proper Motions Exceeding 0."5 Annually, Willem J. Luyten, 1955, The Lund Press, Minneapolis, Minnesota

Catalogue of Stars within Twenty-Five Parsecs of the Sun, Sir Richard Woolley, Elizabeth A. Epps, Margaret J. Penston, and Susan B. Pocock, Herstmonceux: Royal Greenwich Observatory, 1970, Joint Publications of the Royal Greenwich and Cape Observatories

Constellation Star Game – 36 Constellations of Stars, Vinson Brown, Revised edition, 1974, Naturegraph Publishers, Inc., Happy Camp, California

Cycle of Celestial Objects, Vol. 2 – The Bedford Catalogue, Captain William Henry Smyth, First edition, 1844, John W. Parker, London

Délimitation Scientifique des Constellations, Report of Commission 3 of the International Astronomical Union, Eugène Delporte, First edition, 1930, Cambridge University Press

The Exploration of the Universe, George O. Abell, Sixth edition, 1991, Saunders College Publishing, Philadelphia

The Facts on File Dictionary of Astronomy, Valerie Illingworth, editor, Second edition, 1985, Facts on File Publications, New York

The Glorious Constellations – History and Mythology, Giuseppe Maria Sesti, First English translation edition, 1991, Harry N. Abrams, Inc., New York

Handbook of the Constellations, Hans Vehrenberg and Dieter Blank, Second edition, 1973, Treugesell-Verlag, Düsseldorf

Maps of the Heavens, George Sergeant Snyder, First edition, 1984, Andre Deutsch, Ltd, London

The Messier Album, John H. Mallas and Evered Kreimer, First edition, 1978, Sky Publishing Corporation, Cambridge, Massachusetts

Naked Eye Stars – Catalogued by Constellation and in Three Groups by Brightness, Richard H. Lampkin, Second edition, 1972, Gall and Inglis, Edinburgh

Norton's Star Atlas, Arthur P. Norton, Sixteenth edition, 1973, Sky Publishing Corporation, Cambridge, Massachusetts

Observing Handbook and Catalogue of Deep-Sky Objects, Christian B. Luginbuhl and Brian A. Skiff, First edition, 1990, Cambridge University Press

The Sky Explored – Celestial Cartography 1500–1800, Deborah J. Warner, First edition, 1979, Alan R. Liss, Inc., New York

The Star Guide, Steven L. Beyer, First edition, 1986, Little, Brown, & Co., Boston

Star Names and Their Meanings, Richard Hinckley Allen, First edition, 1899,
G. E. Stechert, New York and London, republished as *Star Names: Their Lore and Meaning,* 1963 Dover Publications, New York

Universe Guide to Stars and Planets, Ian Ridpath and Wil Tirion, First edition, 1985, Universe Books, New York

A working list of meteor streams, Allan F. Cook, Evolutionary and Physical Properties of Meteorites, IAU Colloquium No. 13, NASA SP-319, C. L. Hemenway, P. M. Millman, and A. F. Cook, editors, 1973

World Wide Planisphere, William H. Barton, Jr, First edition, 1943, Addison-Wesley Press, Cambridge, Massachusetts